T0100271

SELF-DISCIPLINE AND NATURE

www.royalcollins.com

SELF-DISCIPLINE AND NATURE

A HISTORY OF HUMANITY'S WAR AGAINST THE PANDEMIC

KEVIN CHEN

Self-discipline and Nature: A History of Humanity's War Against the Pandemic

KEVIN CHEN

First published in 2022 by Royal Collins Publishing Group Inc.
Groupe Publication Royal Collins Inc.
BKM Royalcollins Publishers Private Limited

Headquarters: 550-555 boul. René-Lévesque O Montréal (Québec) H2Z1B1 Canada
India office: 805 Hemkunt House, 8th Floor, Rajendra Place, New Delhi 110 008

ISBN: 978-1-4878-0916-4

To find out more about our publications, please visit www.royalcollins.com.

Preface

Since coronavirus has spread across the globe, it became the most important variable affecting the global economy, politics, and other aspects of our lives. The pandemic has resulted in increasing social uncertainty, with different issues emerging in various fields. It has also posed great challenges from education to medical care and even social ideology. Its expanding scope of infection, as well as its continuity, has recalibrated our expectations and further aggravated social vulnerability.

The lockdown due to the pandemic has resulted in the largest increase in unemployment levels since 1930. According to the US Department of Labor Statistics in April 2020, as of 11 April, the percentage of successful applicants for unemployment benefits rose to a record 11% of the total labor force. Concurrently, the number of applications for unemployment benefits five weeks since lockdown climbed to 26 million, surpassing the total jobs created in the United States since the financial crisis.

The rapid economic downturn shocked the world. According to China's GDP report for the first quarter of 2020, the primary industry was the least impacted at –3.2%, while the secondary industry was the most impacted at –9.6%. This was due to labor movement restrictions, thus affecting production lines and sending shockwaves. The next most impacted industry was the tertiary industry at –5.2%.

The pandemic has also created more variables for international relations, which have been volatile in recent years. In comparison with the response to the financial crisis, the effectiveness of international cooperation efforts for the pandemic was lackluster. The World Health

Organization (WHO) was at a loss, while the United States lambasted the WHO, strongly demanding to halt its funding and threatening to never pay its annual dues in the future. The differences in principles between major nations made it challenging for the G20 meeting to replicate its past efficiency and determination. Countries were acting at their own will, China and the United States confronted each other, as well as other sovereign governments, who have gradually retracted the power that they once had transferred to the international mechanisms.

The pandemic across the world has turned the world outside our homes into a wilderness and decreased the public's engagement in public spaces due to social distancing. It has also resulted in short-term emergency measures that became a part of our social lives. This is the nature of such measures, which have accelerated their historical process. Decisions, which took several years to pass under normal circumstances, were now made within a few hours. From an international space to individual countries, everyone was a guinea pig in a large-scale social experiment.

The pandemic is not a temporal disruption to society. It is the beginning of a completely different way of life.

Some countries identify persons at risk of contracting diseases through methods with greater precision. Israel tracks those who are in contact with known virus carriers using the location data of terrorists' mobile phones. Singapore has conducted detailed contact tracing by publishing the details of known cases, including where the patients work and live, as well as the hospitals that they were admitted to.

MIT Technology Review has imagined the impact of the pandemic. In the new world, to board a flight, we must register for a service allowing for our movements to be tracked via mobile phones. While the airlines are unable to gain access to the information, mobile phones will issue an alert when we are close to an infected person or a disease hotspot. Similar requirements will also apply to the entrances of large venues, government buildings, or public transportation hubs. There will be temperature scanners everywhere, and workplaces may require everyone to wear a monitoring device to track their temperature and vital signs. We will eventually adapt to and accept these measures, akin to how we adapt to increasingly stringent airport security checks after terrorist attacks.

Undoubtedly, modern society has made great progress in managing diseases. Previous pandemics have reduced life expectancies for a prolonged period. While advances in medical standards and technological advances have improved the situation, society's fragility remains in the face of a pandemic.

It is almost certain that the coronavirus, just like any plague in history, will have a profound impact on social production and lives. Although it has created a unique, depressing situation with weak demand and supply, as well as geopolitical shocks, it has also accelerated a challenging transformation in global economic, political, and social governance.

Throughout our history, global major accidents were in reality the norm, and they continue to occur throughout history. From the Black Death to the two World Wars, the course of history constantly changes with these accidents, with some of them marking the beginning of a new era.

In the past 30 years, our society, under an unprecedented period of peace, has been developing steadily. Although accidents and challenges continue to occur locally, just like how the coronavirus disrupted our once linear, smooth and predictable society, no single global event could stop this process.

This book was written during the COVID-19 pandemic. With this in mind, it reviewed plagues such as leprosy and smallpox, as well as provided classification and explanation on new viruses such as SARS and Ebola based on modern medical research findings. With pandemic as its main theme and changes serving as clues, it provides observations on the medical, economic, political, and social aspects affecting us during the pandemic, and puts forth a forward-looking, international perspective of the post-pandemic era. The book is interesting and has a scientific basis. Apart from providing more information about the post-pandemic era, it helps readers to understand more about these subjects and to appreciate how the pandemic has been moving forward indefinitely, alongside mankind and civilization.

KEVIN CHEN

Contents

1

Plague Chronicles

1.1 Mankind and Civilization, Mankind and Plague

About 3.8 billion years ago, the Big Bang occurred. This was the beginning of all historical dates and the creation of the world.

The Earth emerged 4.6 billion years ago. Then, 600 million years later, the earliest life forms appeared in oceans during their early stages, with organisms beginning their complex and long evolution from prokaryotes to eukaryotes.

Soon enough, 20 million years ago, some primates began spending more time living on land. About 7 million years ago, the first "man-like apes," which stood on two legs, appeared somewhere in Africa.

Two million years ago, other humanoid species appeared in eastern Africa, also known as the "*Homo habilis.*" What is special about them was that their members could make simple stone tools. They began improving stone tools in their hands and even attempted taming fiery flames. With the combined effects of natural selection and genetic mutations, the brain capacity of their offspring became larger until the *Homo erectus* emerged.

According to paleontological studies, the size and brain capacity of the *Homo erectus* species was similar to humans. The stone tools they made were more precise and complicated than those made by humans. Thereafter, part of the species left Africa and undergone many generations of procreation and migration, which affected the way humans appear in many countries and regions.

Finally, modern humans, known as *homo sapiens*, appeared in East Africa about 250,000 years ago.

Mankind's evolution on Earth has been a long-standing affair. During these tens of millions of years, mankind has been growing together with everything around us, including viruses and plagues that have been accompanying mankind for thousands of years.

The above perspective was based on data from archaeology and its related disciplines and does not represent the true origin of human beings. Perhaps, Africa may just be an inference point based on mankind's current findings and not the actual starting point. However, this inference was put forth from the perspective of evolution. From the perspectives of religion and ethnicity, the origin of mankind was another story, because these creationist perspectives were completely different from the evolution theory. The theory of the origin of mankind is not the focus of this book, while the theory on mankind's evolution is skewed towards Darwin's academic theory. As for myself, I am geared towards the creationism perspective on this issue.

1.1.1 A Virus Planet

Viruses have affected the development of human well-being. Looking back, when did viruses appear on Earth and begin having a connection and relationship with humans?

In the book titled *A Minimal History of Mankind*, the author David Christian described and reviewed mankind's position in the universe. If the entire 13-billion-year history of the universe's evolution is simplified to 13 years, then mankind emerged approximately three days ago, the earliest agricultural civilization occurred five minutes ago, the industrial revolution occurred only six seconds ago, while the world's population reached 6 billion. Both the Second World War and the Apollo 11 moon landing only lasted one second each.

Thus, we can then see mankind's position in the universe.

Inferring from the current archaeological point of view, the Earth is about 4.6 billion years old, while single-celled organisms appeared approximately 3 billion years ago, equivalent to 3 years ago in accordance with the *Minimalist Human History* conversion. In this regard, viruses seemed to exist longer than humans, which also meant that they existed when the first cell was born on Earth, making their history perhaps longer than our history.

Viruses shaped the Earth's environment of today, and they can be found in every corner of the Earth. They exist in every place that people can imagine, such as oceans, glaciers, deserts, volcanoes and of course in human bodies, and leave a host of information in our DNA. Eight percent of human DNA fragments originated from viruses. Hence, it is not an exaggeration to say that they are our ancestors, which are somewhat distant from humans.

The ocean is the place with the most viruses on Earth. In a liter of seawater, there are about 100 billion viruses. These viruses contribute to the evolution of marine life while undergoing gene exchange. It can be said that the virus maintains the ecological balance and silently changes the Earth's environment.

The word "virus" has undergone three rounds of changes before it became closer to its current meaning. The word "virus" is based on a Latin word for "poisonous secretion," and by extension, it means "snake venom" and "human semen," thus giving it a dual meaning of "destruction" and "creation." Over the centuries, it gradually took on another meaning – infectious substances, such as pus from wounds and mysterious substances spreading through the air.

In the late 19th century, when researching the causes of tobacco mosaic virus, Ivanovsky named it the "filtering virus" as its pathogenic factor did not match a bacteria's characteristics. This laid the foundation of virology, and that was also the first time that humans used the word "virus" in this context. As the first person to discover a virus in the world, he is also known as the "father of virology" by future generations.

In the 1920s, Stanley concentrated and separated protein crystals from the extract of tobacco leaves infected with the tobacco mosaic virus. Experiments proved that these crystals were infectious and can proliferate themselves. They also discovered the structure of the virus, which is 95% protein + 5% nucleic acid.

Through the assistance of Halmet Ruska, the younger brother of Ernst Ruska who was the inventor of the electron microscope and 1986 Nobel Prize winner in Physics, German biochemist Robert Koch finally observed the tobacco mosaic virus in 1939 and confirmed that it was a rod-shaped particle. From the initial presumption of the tobacco mosaic virus as a filtering pathogen to the direct observation of this filtering pathogen as a submicroscopic particle, it took mankind a total of 41 years.

The in-depth study of viruses was enabled by modern medicine's development. According to its definition, a virus has a small, simple structure, containing only one type of nucleic acid (DNA or RNA), is parasitic in living cells, and multiplies by replication.

The virus exists in a non-cellular life form, comprising a long non-coding RNA and a capsid. The virus does not have its own metabolic mechanism and enzyme system. Therefore, when the virus leaves the host cell, it becomes a chemical substance that is lifeless and cannot reproduce independently. Its replication, transcription, and translation capabilities are carried out in the host cell. When it enters the host cell, it can use the material and energy in the cell to complete its life activities and produce a new generation of similar viruses based on the genetic information contained in its nucleic acid.

Viruses are not always a menace. The oxygen that humans inhale comprises 1/10th of viruses, and most viruses are beneficial to the human body. For example, viruses can help the human body balance the bacterial community in the intestines to prevent damage to human bodies. The viruses, which shuttle between different hosts, have some probability of carrying the gene fragments of the previous host and inserting themselves into the genes of their next host, thus enabling the diversity of humans or other species.

Of course, viruses are also cold-blooded killers created by nature. At present, there are 320,000 viruses, which are deadly to mammals, and their harm to humans is unquestionable. As most of the famous viruses, such as smallpox, AIDS, and Ebola viruses are fatal, people are fearful to discover them.

Most existing viruses cannot perform metabolic activities on their own and need to "reproduce" by injecting nucleic acids into host cells. Once viral nucleic acid enters the host cell's nucleus to proliferate, it will disrupt its usual translation and transcription. The invasion of host cells by antigen or over-immunity arising from virus invasion would cause damage or even death of the host cells, resulting in an irreversible negative effect on human health.

1.1.2 Plague Civilization

The emergence of human civilization occurred within the last 10,000 years, and another big change that occurred almost concurrently was the formation of infectious diseases. With the development of civilization, infectious diseases have also developed at a similar or even faster speed, with cross-regional, cross-continental, and global epidemics occurring at or around the AD period.

Infectious diseases are also called plagues. The emergence of plagues is due to population growth, human settlement, agriculture, and animal husbandry. It can be said that it was an inevitable consequence of globalization that happened with the wave of civilization. During the last thousands of years, the plague has always coexisted with mankind, affecting all aspects of society such as politics, economy, and culture.

The earliest and most detailed narrative about the plague can be found in the first volume of Homer's Iliad published in the 6th century BC. Brad described the plague as such: "Achilles and Agamemnon quarreled and began a feud. Lifting one's voice, Goddess! The fury of Achilles, son of Peleus was unleashed... Which God incited the feud between them? It was Apollo, son of Zeus and Leto. He was dissatisfied with the king and caused a deadly plague in his army, devouring the lives of his soldiers."

In *Homer's Epic*, the poet's description of the plague only provided us with a superficial view of the narration and struggle against the plague by people living in ancient times. Although the book *The Plague Was Used by Apollo To Punish the World* is a metaphor for reality, it cannot be taken into consideration during the plague.

From 430 to 427 BC, the plague, which broke out in the city-state of Athens and destroyed the glorious city-state civilization, had a more detailed record. "Physically fit people were suddenly hit by severe high fevers, with eyes turning fiery red as if flames were ejected. Their throats or tongues bled and emitted an unnatural stench, alongside vomiting and diarrhea accompanied by extreme thirst. By then, their bodies were in pain and inflamed with ulcers, unable to sleep or tolerate the touch of the bed. Some patients were naked on the street wandering while looking for water for a sip until they fell to the ground and died. Even dogs fell victim to this disease, alongside crows and eagles, who ate the corpses of the people lying everywhere. The people who survived either lost their fingers, toes, and eyes or lost their memory", said Thucydides, a Greek historian while describing the plague that destroyed Athens.

When the plague broke out, it was during the Peloponnesian War, the largest civil war in ancient Greece, when Athens and Sparta were engaged in a fierce military confrontation. The spread of the plague wiped out a quarter of the population. Even Pericles, the Chief General, was not spared. According to historical records, it originated in Africa, made its way from Ethiopia via sea to the port of Piraeus, before reaching the Acropolis. His death had a direct impact on the Peloponnesian War. In 404 BC, Athens was defeated and surrendered to Sparta, and ancient Greece was no more. For the next two thousand years, there was never a glorious city-state civilization like Athens in Europe again.

The largest plague outbreak in history, known as the Antonine Plague, occurred in 168 A.D. Nicolas Poussin faithfully recorded the situation in his painting The Plague of Ashdod. "The corpses cracked and decomposed on the street because they were not buried. Their abdomens were swollen, with their mouths wide open and oozing pus like a torrent. Their eyes were flushed, and hands were raised. The corpses were stacked on top of one another and rotting in corners, streets, courtyard porches, and churches."

What kind of disease was this? Scholars of the past generations had put forward different explanations, including smallpox, plague, flu, and cholera. The plague lasted approximately fifteen years and wiped out one-third of the Roman Empire's population. After the plague, the golden age of Rome was over. The Empire's society and politics, especially literature and art, suffered a huge blow. The pessimism, despair, and sudden death of King Antonius, a famous poet, and philosopher at the time, had a devasting effect on the cultural confidence of Roman society. Also, those who were in despair resorted to witchcraft and shifted their belief in God, thus formed a vicious cycle and led to the eventual collapse of the Roman Empire.

Plague refers to a pandemic, too. Many people were infected and sick with the same disease. During the thousands of years of human existence, the plague had destroyed and created civilizations.

1.2 Ancient Leprosy and Smallpox

In religion, humans believed that a secret connection existed between the human spirit and the gods, and tended to establish their supremacy in the ecosphere. Christians believed that God created Adam with clay, Buddhists attributed the existence and reproduction of humans to the boundless cycle of death and rebirth, while other religions regarded mankind as a mystical creation of God.

However in reality, regardless of which notion humans believe, the body outside the spirit can only be accompanied by cells and viruses. Evolution over millions of years has enabled humans to possess extraordinary skills, but they are still unable to manage the evasion of the body from the world and are constantly fighting against endless types and quantities of bacteria or viruses to continue their lives of limited existence.

When a type of bacteria or virus attacks a certain living body and makes it suffer, this is an individual disease. However, when some fatal and invasive bacteria or viruses attack humans, the disease can result in massive damage and deaths, while it spreads across a large area. In this instance, the individual disease would become a frightening plague.

Among them, leprosy and smallpox have been two of the oldest plagues accompanying humans.

1.2.1 The Leprosy Nightmare

It can be said that leprosy emerged at almost the same time as human civilization. Leprosy has a long history on Earth, and its spread has caused its patients to exist in almost every country and region across five continents.

The records of leprosy can be traced to ancient Egypt. According to textual research of the papyrus in 2400 BC, the word "set" refers to leprosy. The dolium, whose production period was equivalent to 1411–1314 BC, was found among the fourth-generation pharaoh of Egypt's palace remains bearing an engraving similar to the "facies leonine" of leprosy.

On the other hand, Indian scholars, who referred to the word "Kushtha" in Vedas, a Vedic Sanskrit in 1400 BC, believed that leprosy was prevalent in India for at least 3000 years. In addition, cuneiforms, including legal provisions to keep leprosy patients away from the city, were discovered at the ruins of Ashurbanipal Palace in Nineveh, which was built in the 7th century BC and one of the previously largest cities in the world. This meant that leprosy was already prevalent at the Tigris and Euphrates river basins of western Asia.

The earliest leprosy-related records in China were during the Yin and Shang Dynasties (1066 BC) where Qi Zi painted his body to replicate leprosy-like conditions and escaped death as cited in the *Annals of the Warring States* (*zhanguoche*). In fact, there were already many leprosy records recorded in Chinese medicine classics after the Qin and Han Dynasties, such as the Warring States period's *Inner Canon*, Sui Dynasty's *General Treatise on Causes and Manifestations of all Diseases*, Tang Dynasty's *Three Volumes on Gynaecology*, the Qing Dynasty's *Jiewei Wuxuan* and *Complete Collection of Wind Diseases*, etc. These classics already had a deeper understanding and systematic discussion on the symptoms and treatment of leprosy.

In the *New Testament* of Christianity, there was also a record of Jesus' exposure to leprosy and treatment, but this was after A.D. However, what we know is that leprosy is an ancient and long-standing infectious disease.

Leprosy in Europe was a product of the Crusades (1096–1291). During this large-scale military expansion, King Baldwin IV of Jerusalem received special attention because of his leprosy patient status. The "Leprosy King" was unable to govern the country due to illness, which led to the decline of the Kingdom of Jerusalem.

Leprosy has plagued mankind for more than 3000 years, alongside the struggle between both sides.

Modern medicine revealed the mechanism of leprosy's transmission. Also known as the "Hansen disease," it is a chronic infectious disease caused by Mycobacterium leprae, affecting mainly the skin and peripheral nerves. It is named after Norwegian scholar Gerhard Armauer Hansen, who discovered it in 1873.

The clinical symptoms of leprosy are numbness in the area of skin lesions, nerve thickening, and disability in severe cases. If left untreated, it may cause muscle weakness, disfigurement, and permanent nerve damage. According to the body's immune status, case changes, and clinical diagnosis, most patients can be divided into lepromatous and tuberculoid-like types. A small number of patients are under the borderline category and other unspecified categories. Lepromatous and tuberculoid leprosy patients are the most infectious.

Leprosy patients are the natural hosts of Mycobacterium leprae, and the source of leprosy infections is untreated leprosy patients. Among them, the skin and mucous membranes of multi-bacteria patients containing large numbers of Mycobacterium leprae are key infection sources. When an infected person coughs or sneezes, the disease will be spread through droplets, broken skin, and mucous membranes and close contact, etc. The susceptibility of humans to Mycobacterium leprae is very inconsistent. Children are at a higher risk of developing leprosy than adults, while most of the cases are adults aged above 20. There are also more male cases than female cases.

In the early 1940s, after sulfone drugs were proven effective in treating leprosy, the age of chemotherapy arrived. Leprosy patients were no longer helpless and awaiting death. Today, leprosy is no longer a cause for alarm in the medical world. Multiple drugs combined with chemotherapy can completely cure leprosy within six to twelve months. Early detection and timely treatment can avoid any disability, while leprosy patients who were cured are no longer infectious. With the increasing number of cured cases and the normalization of prevention and treatment measures, the eradication of leprosy worldwide is nearing.

1.2.2 History of the Death of Smallpox

On 25 October 1977, the last smallpox patient was reported in Somalia, Africa. No smallpox patient was discovered thereafter. Smallpox, one of the deadliest epidemics in human history, had since disappeared. This moment came earlier in China. With the recovery of the last patient in 1961, no further smallpox cases were reported in China.

In October 1979, the WHO confirmed that China had no smallpox cases since the 1960s. In May 1980, the 33rd World Health Assembly officially announced that humans had eliminated smallpox. So far, smallpox was the only infectious disease eliminated by humans through their efforts and scientific methods.

Smallpox was one of the oldest known infectious diseases. Pharaoh Ramses V of ancient Egypt was the first recorded smallpox case. From his death in 1145 BC to 1980, where smallpox was declared eradicated, this infectious disease, which was the fear of mankind, had raged on for more than 3,000 years.

Specifically, smallpox was a severe infectious disease caused by the smallpox virus. The mortality rate of variola major was above 25%, while the mortality rate of hemorrhagic smallpox reached 97%. When smallpox occurs, the patient has skin sores filled with a thick, opaque fluid all over their body, hence this plague is named "smallpox."

Everyone, regardless of age, was vulnerable to smallpox. It could be transmitted through the inhalation of droplets or direct contact with patients. Among them, the saliva of patients, who had been down with the virus for a week was the most infectious, because it contained the highest levels of variola virus. Smallpox was also a DNA virus, which was highly infectious and had a high mortality rate, and it was described as a "sniper" aimed at humans. During ancient times, once infected with smallpox, there was almost no cure apart from self-immunity and taking medication to relieve symptoms. The typical symptoms included chills, high fever, fatigue, headache, leg, and lower back pain, as well as typhus, pimples, herpes, and pustules. If the body temperature rose sharply, it could cause convulsions and coma.

In addition, patients with severe smallpox often have many complications, such as sepsis, encephalitis, osteomyelitis, and pneumonia, etc. The majority of these complications can be serious, and the main causes of death arising from smallpox. The people who survived would have many scars due to scabbing, known as a "pockmarked face."

In the era of poor medical standards and sanitation, everyone, including the emperor, nobles, and ordinary citizens, faced the threat of death due to smallpox. According to records, four emperors from the Qing Dynasty suffered from smallpox, of which Emperors Shunzhi and Tongzhi died it, while Emperors Kangxi and Xianfeng, who escaped death, had pockmarks on their faces.

The earliest smallpox cases were recorded in China in the Eastern Jin Dynasty. Smallpox symptoms were described in Medical scientist Ge Hong's (283–363) book *The Handbook of Prescriptions for Emergencies*. However, medical personnel at that time did not know enough about the smallpox virus, and the book did not provide a specific treatment method.

As time passed, humans finally discovered its weakness in subsequent treatments. Surviving a smallpox infection would lead to lifelong immunity. This characteristic inspired ancient physicians to "attack poison with poison," attempting the use of "paternal blood," "pus" or direct contact with mildly infected patients to be infected to gain immunity. These attempts achieved some results.

On that basis, a "human pox" technique, which began in the Song Dynasty and matured during the Ming and Qing Dynasties, was devised. Simply put, dry power made from a patient's scab would be blown into the nose using a tube, with the hope of making the recipient immune to the smallpox virus. Although many patients died from this technique, others survived. Gradually, this method became widely used. It can be said that it ushered in an era of human immunotherapy.

The news of China achieving results in the prevention and control of smallpox quickly spread to other parts of the world. It brought hope to all who were suffering from the smallpox virus. However, although the technique can indeed prevent smallpox, the possibility of death remained, so it was not considered an ideal cure. It was not until the eighteenth century in Europe that a British doctor named Edward Jenner invented the vaccination method.

Jenner observed that there would be many pockmarks on the face after contracting smallpox. Many European women used powder to conceal these scars. But there was a category of women who neither wore makeup nor had acne marks on their faces. They were milkmaids on the farm. In general, they do not get acne or develop pockmarks. On 14 May 1796, he found a young milkmaid who contracted cowpox recently, collected pus from her skin, and injected it into Phips, an 8-year-old boy, who recovered after a few days of fever. Two months later, Jenner vaccinated him again, this time using pus taken from a smallpox patient's wounds. Phipps showed no symptoms.

This strengthened his confidence in smallpox vaccination, which was the prototype of the vaccine, and enabled him to continue its development. In fact, during the early stage of research and development, some Europeans did not trust it and even mocked Jenner and vaccinators. However, people saw its benefits after the untiring efforts by a large number of doctors led by Jenner himself to promote its usage. Since then, vaccination had replaced the human pox technique and became the primary choice by humans to prevent smallpox. After the mass production, this method became common in European countries and their colonies and led to the demise of smallpox.

1.3 Black Death — Plague

In the development of human history, the Black Death, Spanish Flu, Smallpox, and AIDS were named the "Four Plagues." Among them, the Black Death originated from Africa. According to the papyrus literature records at "Dr. Rajab Papyrus Museum," an infectious plague of hemorrhagic fever occurred between 1500 BC and 1350 BC in ancient Egypt at the Nile Valley, which was located on the banks of the Nile River in Cairo, Egypt. It was later named the Black Death.

The Black Death, named after corpses that turned dark purple, originated from a terrifying plague in the Middle Ages. It was confirmed to be a plague by science.

1.3.1 From Renaissance to the End of Dynasties

Plague, caused by the bacteria Yersinia pestis, is an international quarantinable infectious disease. The *Law of the People's Republic of China on the Prevention and Control of Infectious Diseases* divides infectious diseases into three categories, namely A, B, and C according to their severity and harm to the public. Plague, undisputedly the number one killer, alongside cholera are listed under Class A.

As an infectious disease originating from nature, plague is prevalent among rodents, with rats and marmots being the more common natural hosts of Yersinia pestis. In addition, wild foxes, wild wolves, wild cats, hares, camels, and sheep may be infected with the plague and become infection sources. Since natural hosts are mostly rodents, in a plague natural foci, vector organisms are the main transmission routes. Through fleas, Yersinia pestis can be transmitted from the infection source to people.

The incubation period of plague is relatively short, generally between 1–6 days, mostly 2–3 days, and between 8–9 days in some cases. According to different infection areas and clinical conditions, the plague can be divided into the bubonic plague, pneumonic plague, septicemic plague, mild plague, and other rare types of plague.

Glandular type is the most common and often occurs in the early stages of a plague, with sudden chills and high fever, headache, and fatigue, body aches, occasional nausea and vomiting, irritability, ecchymosis, and bleeding. The "Black Death" of the Middle Ages is the bubonic plague. The pneumonic plague may originate directly or indirectly from the lungs and develop rapidly, causing a rapid onset of high fever, with obvious systems of systemic poisoning. After being infected for a few hours, chest pains, cough, sputum, which begins with a small amount rapidly becomes frothy, blood-streaked sputum. The septicemic plague type develops systemic toxemia and central nervous system symptoms, as well as rapidly develops into severe bleeding. The typical symptoms from mild patients include irregular fever, mild systemic symptoms, swelling and pain in the lymph nodes and no bleeding, are more common in the early and late stages of epidemics or those who are vaccinated.

Although the plague is terrifying, developments in public health and medical technology have ensured its prevention, control, and treatment. Since the discovery of Yersinia pestis, the culprit of the plague in 1894, people could quickly test whether they were down with the plague through nucleic acid and pathogenic tests, as well as other means. Early diagnosis, as well as the timely use of various antibacterial drugs, greatly reduced its fatality rate. Stringent isolation conditions also enabled the infection to be well-controlled.

Since rodents are its natural hosts, people can't eliminate rodents on Earth, so it is difficult for the plague to be eliminated, just like smallpox. Although the plague occurs occasionally, it is difficult for modern medicine to create a large-scale impact.

The plague, as a severe infectious disease, has extensively and profoundly affected the progress of human civilization.

Plague: The Han Empire Crisis

According to historical records, between the end of the Eastern Han Dynasty and the start of the Three Kingdoms, an epidemic occurred in the Han Empire and reached its peak towards the end of the Eastern Han Dynasty. The last thirty years of the Eastern Han Dynasty, between 204–219, were known as one of the most terrifying years in Chinese history as the plague's fatality rate exceeded 60%. As described by Zhang Zhongjing in *Treatise on Cold Damage Diseases and Deaths*, "In less than a decade, more than two-thirds of the population, including 200 of my family members, were wiped out." China's population, which plummeted from more than 60 million to less than 15 million, alongside deaths forming two-thirds of the total population, also recorded its largest population decline in history. And which infectious disease caused the great plague during the late Eastern Han Dynasty? There is no clear answer. At present, there are various theories such as the plague, epidemic hemorrhagic fever, typhoid fever, and influenza, while some thought it was an ancient infectious disease that has disappeared. Based on the symptoms of most victims, it was an infectious disease spread by mammals acting as the virus-host and it was characterized by sudden high fever and high fatality rate. Thus, it was inferred that the plague is the most likely disease. This is why most currently believe that the end of the Eastern Han Dynasty was the origin of the first plague.

Plague of Justinian

Western history has a clear record of plagues. The first plague, which attacked humans, occurred in the 6th century. To be precise, it began in 520 AD, although it was also said that it started in 541 AD. The epidemic originated from Egypt's Sinai Peninsula and later spread to the entire Mediterranean region for 50 to 60 years. According to records, during the worst period, 50,000 to 10,000 people were dying every day, with the total number of deaths exceeding 100 million worldwide, causing the European population to halve.

During this period, as the most severely affected Ethiopian region coincided with the reign of the Justinian dynasty, it was called the Plague of Justinian or Justinianic plague. It is the worldwide plague well documented in medical history, which spread across the entire Mediterranean coast. Its direct impact was the Eastern Roman Empire's rapid decline, the Arab Empire's rapid rise, and the Islamic civilization's expansion to Europe.

Black Death Gave Rise to the Renaissance

After the Plague of Justinian, the plague seemed to have taken a step back while continuing to lurk behind the scenes. However, when it re-emerged, mankind discovered how fragile its civilization is.

In the 14th century, the plague launched its second large-scale attack on mankind in the vast Gobi of Central Asia and followed the Mongolian army in its western expedition. In the summer of 1346, Genea's colony of Kaffa (modern Feodosiya) was surrounded by Tatars, who threw infected corpses into the city and caused the plague to spread in the city.

From 1347 to 1353, a major plague swept across Europe. As those infected had herpes on their skin and black scabs were appearing on their skin when it was torn, with the skin sores turning black in severe cases. This was known as the "Black Death." It claimed the lives of 25 million Europeans, which is one-third of the European population at that time. It later spread to Russia, wiping out nearly half of its population.

Since then, in the 15th and 16th centuries, the Black Death returned and ravaged the European continent, causing 50 to 75 million deaths. Its worst resurgence is probably the outbreak that occurred in London from 1665 to 1666 and killed 75,000 to 100,000 people in London, which exceeded 20% of its total population at the time.

This disaster also triggered crises in Western politics, culture, economy, religion, and social structure, which in turn triggered a series of profound social changes. It can be said that this plague directly gave birth to contemporary Western civilization. Boccaccio's famous book *The*

Decameron told the story of the European Black Death pandemic. This work full of humanistic ideas was considered a manifesto of the European Renaissance. Many people in modern times believe that it was the epidemic of the Black Death that shook the authority of European religious forces, promoted the development of medicine and science, and enabled Europe to emerge from the dark Middle Ages and enter the Renaissance period.

The Plague Ended the Ming Dynasty

During Wanli of the Ming Dynasty, there were more droughts, alongside plagues. The most serious plague began in the 9th year of Chongzhen (1636) when Li Zicheng defeated Ming Yulin's army in Shaanxi Anding. "There was widespread famine and plague was severe, and it was most serious at Waisaibao." In the 10th year of Chongzhen, Zhang Xianzhong captured Qichun and Huangshi in Hubei, ravaged Jiujiang in Jiangxi, resulting in famine and pandemics in these areas. In the 11th year of Chongzhen, the pandemic at Taiping Prefecture, Nanzhili province caused many deaths. In the 12th year of Chongzhen, there was a severe drought in Licheng, Qihe, Shandong, followed by a pandemic. It was recorded in history as a severe plague with an extremely high fatality rate. Those who were infected died quickly.

At the end of the Ming Dynasty, the population of Beijing was about 800,000 to 1 million. At the plague's peak, there were 10,000 deaths daily. At the end of Chongzhen's 16th year of reign, 20% of the population died in the capital city. Resources were scarce, prices were soaring, and the people were living a hard life.

The people were mourning, alongside people lying in the fields after they starved to death. At the end of the Ming Dynasty, Shanhaiguan's population was about 100 million people. Unnatural deaths due to plagues, catastrophes, and wars accounted for about 40% of the population, including 10 million people dying from the plague in North China.

The plague also greatly undermined the Ming's army combat effectiveness. As it raged, the Ming army only had 50,000 soldiers, who were all skin and bones. Many were unable to fight due to insufficient rations and starvation. On average, in the capital of 15,000 – 16,000 crenelations, one soldier was guarding about three crenels. How would they be able to resist Li Zicheng's 500,000 elite soldiers?

The changes in people's hearts caused by the plague were more terrifying than the plague itself. The dynasty's inability resulted in them giving up on it psychologically. Although Chongzhen was diligent, he was unable to save the Ming dynasty, which became unstable due to politicking by eunuchs from the previous dynasties. Ming Dynasty was over, its fate was cast in stone.

1.3.2 A Metaphor of Human Civilization

In 1855, a large-scale plague began in Yunnan, China. It coincided with an eventual period, comprising Du Wenxiu's uprising in Yunnan, and a plague that was spreading in the crowds. In 1894, a plague broke out in Guangdong. It spread across the city within ten days and reached Hong Kong, China. These places became the epicenters exacerbated by the convenience of sea transportation. It eventually extended its reach across the globe, causing approximately 12 million deaths in China and India alone (note: According to Cao Shuji (2005), the plague occurred in Guangzhou, Hong Kong, and Shanghai in 1894, as published in the *Journal of Shanghai Jiaotong University* and collated by *Shen Bao* reports during that period).

However, unlike previous plagues, mankind used modern medicine as a bargaining chip in their fight against death.

In 1894, Alexander Yersing, a bacteriologist from the Pasteur Institute, isolated the plague bacillus from plague patients in Hong Kong, China. In 1898, French scientist Paul Louis Simond proved for the first time in Mumbai, India that rats and fleas were the primary culprits in the spread of plague, which also marked the beginning of the fight against the thousand-year plague using modern medicine.

In 1910, where the Qing Dynasty was still under the second year of Emperor Xuantong's reign and a year before the Chinese Revolution of 1911, the three eastern provinces were still under the sphere of influence of both, Japan and Russia. In this context, the plague broke out in the Northeast, especially in the Fujiadian (now Daowai District) area of Harbin.

Although historical records enabled the Qing rulers to be very clear about the plague's dangers, the dynasty was already on the verge of collapsing and it was losing its ability to rule in the second year of Xuantong. Since the first report of a case occurring in Manzhouli on 26 October, emergency reports were sent one after another from the Northeast. By 15 November, the Fujiadian area of Harbin was overwhelmed with corpses, and it was too late for them to be buried. The Qing government took no further action, apart from sending troops to barricade Shanhaiguan to prevent outsiders from entering.

Concurrently, Tsarist Russia and Japan also wanted to take advantage of the situation by seizing the Northeast's plague control rights in the region. Under such circumstances, Wu Lien-teh was appointed the Chief Medical Officer for plague control and was put in charge of medical work in these provinces. He was serving as a Vice Director at the Army Medical College based in Tianjin. It was renamed as National Defense Medical College, then shifted to Shanghai, before moving part of its operations to Taiwan in 1949. It was formerly part of The Second Military Medical University.

On 24 December 1910, Wu Lien-teh's assistant, Lin Jiarui, a senior student at the Army Medical College, reached Harbin, the plague's epicenter. He led the prevention and control efforts and brought it under control within four months. Using modern medical methods, Wu Lien-teh was able to stand his ground and scientifically concluded that the plague was caused by Yersinia pestis. When it was ravaging in Europe, it was only transmittable between mice, fleas, and humans. Conversely, it could also spread from person to person at this time. Due to Wu Lien-teh's accurate assessment and effective measures, a severe infectious disease, which shocked the world and had a death toll of 60,000, was wiped out within four months. It can be said that the efforts were no less than a war.

In early 1911, Wu Lien-teh established China's first plague research institute in Harbin. In April, Dr Wu Lien-teh, Chief Medical Officer for plague control of the three provinces, served as Chairman for the International Plague Research Association, which was held in Fengtian and attended by experts from 11 countries. At the meeting, Chinese and foreign experts suggested that the Qing government set up a permanent plague prevention agency in these provinces to prevent the plague's recurrence. China also regained its sovereign rights over seaport quarantine. In addition, the more compact and practical masks, designed by Wu for epidemic prevention, were well-developed and widely used in later generations and these were considered to be one of the predecessors of N95 masks.

The fight between humans and the plague has never stopped, but it has not been on the right track for a prolonged period. In the early days, people tried to treat the Black Death by swallowing feces and ashes, removing black bumps directly or placing living toads on their chest.

With the discovery of Yersinia pestis, humans had a right understanding of the plague's pathogenesis for the first time. The invention of antibiotics also provided an additional powerful tool for humans' fight against the plague, while the administering of timely treatment for patients suspected of contracting the plague also significantly reduced the mortality rate. Clinical trials proved that streptomycin, gentamicin, tetracycline, fluoroquinolone, or chloramphenicol are equally effective in combating Yersinia pestis.

The fear of the plague resulted in mankind's research into medicine and microbiology, which brought about other prevention and treatment methods in modern medicine. In a sense, the fight against the Black Death accentuated the rise of modern medicine and heightened our awareness of public health issues.

When people discussed the plague, they seemed to be relating an ancient myth, or a strange story passed down from the Middle Ages. However, the plague is not far away from us. It has become a metaphor for the development of human civilization. In reality, things undermining happiness always exist, but they are not visible among the revelers.

Contrary to the perceptions of people from the Middle Ages, modern medicine made us realize that the plague's occurrence is not due to God's punishment or a devil's joke. We have to bear more responsibility for the outbreak of disasters. The pollution and deterioration of the living environment, shortage of medical equipment and staff, failure of social systems, and lack of basic health knowledge are important causes of disasters. The plague only needs to change its form to attack humans repeatedly.

A famous historian Croce once said that all history is contemporary history. The plague's history is also the history of contemporary plague. When humans bit the first apple in the Garden of Eden, humans were destined to face various unpredictable living environments. The plague can also be regarded as one of nature's metabolic methods. Human beings will coexist with plagues, which are existing and have yet to appear, and they may coexist forever.

From a contemporary perspective, we saw the economic and energy crisis, population explosion, and environmental pollution. From a historical perspective, we would see various invisible hands, including the plague manipulating the steering wheel of history. It is only by rebuilding the awe of nature and cultivating the scientific consciousness that we can realize the harmonious symbiosis between nature and us, to survive and develop for a long time.

1.4 The Truth of Influenza

Cold and flu cases will peak when the seasons change. Due to a similarity of early symptoms between cold and flu, as well as inadequate medical science education, most people do not have a clear and scientific understanding of the two and are often confused. However, the cold and flu are different.

1.4.1 Flu and Cold

The rhinovirus, a virus with a spherical structure, is a leading cause of common colds. People have a runny nose when they get a cold because the virus infects our nasal mucosa through the air, resulting in inflammatory reactions and cold symptoms. The virus transmits itself via two ways, namely through droplets and aerosols, as well as contaminated surfaces, including person-to-person contact.

Symptoms of rhinovirus infection include sore throat, runny nose, blocked nose, sneezing, and coughing. Other patients also experience muscle aches, fatigue, headache, muscle weakness and loss of appetite, etc. Conversely, influenza is often associated with fever and extreme fatigue. After infection by the rhinovirus, immunity is gained, albeit for a short time. Also, it offers little protection against the other types of rhinovirus, so people can catch a cold many times. There is no specific prevention and treatment for it.

The flu virus is very different from the rhinovirus, which causes the common cold. Solely on the risk alone, the fatality rate of pneumonia caused by severe influenza is 9%, second only to SARS' fatality rate at 10%, while the common cold results in little or almost no deaths. The flu, known as the influenza virus, is a representative of the Orthomyxoviridae family, including human and animal influenza viruses. The former is divided into three types, namely A, B, and C, which are pathogens of influenza (flu).

The influenza A virus's antigenic drift is prone to mutation and it has caused multiple pandemics globally. For example, the pandemic from 1918 to 1919 resulted in at least 20 to 40 million deaths. The Influenza B virus is primarily a human pathogen, but it was not found to cause worldwide pandemics, while the Influenza C virus will cause either unobvious or mild upper respiratory tract infections in humans, but rarely a pandemic. The influenza A virus was successfully separated in 1933, the influenza B virus in 1940, while the influenza C virus was successfully isolated in 1949.

In terms of vitality, the influenza virus has a relatively weak resistance. It is not heat resistant and can be inactivated by a temperature of 56°C in 30 minutes. Its infectivity is rapidly lost at room temperature, but it can survive for several weeks at 0°C to 4°C or for a long time below −70°C or after freeze-drying. It is also very sensitive to dryness, sunlight, ultraviolet rays, and chemical drugs such as ether, formaldehyde, and lactic acid.

Patients are the main source of infections, followed by those who are asymptomatic. Infected animals may also be a source of infection. It is primarily transmitted via droplets with the influenza virus, which enters the body via the respiratory tract. A small number can also be infected through the sharing of handkerchiefs and towels.

After it was transmitted to humans, it would be highly infectious and it can spread rapidly. Its transmission rate is related to the breadth and population density. If the virus enters the human body that was not cleared by the cough reflex or neutralized by the body's specific IgA antibody and inactivated by non-specific inhibitors in mucus secretions, it can infect a small number of respiratory epithelial cells, causing them to produce vacuoles and degenerate, as well as produce progeny virions, which will spread to neighboring cells and repeat the viral propagation cycle.

The influenza virus will cause degeneration, necrosis, and even shedding of host cells, resulting in mucosa congestion, enema, and increased secretions, which in turn gives rise to nasal congestion, runny nose, sore throat, dry cough, and other upper respiratory tract infection symptoms. When it spreads to the lower respiratory tract, it may cause bronchiolitis and interstitial lung disease.

The population is generally susceptible to the virus. Its incubation period is usually 1 to 4 days, depending on the viral load and body's immune status. The patient will begin displaying symptoms such as chills, headache, fever, body aches, fatigue, nasal congestion, runny nose, sore throat, and cough. Within 1 to 2 days after the onset of these symptoms, the viral load of secretions will be relatively larger before decreasing rapidly thereafter. Asymptomatic patients begin recovering from the 3rd to 4th day. If there are complications, the recovery period will be longer. It is characterized by high morbidity and low mortality, with death usually caused by concurrent bacterial infections, including the common types such as Streptococcus pneumonia, Staphylococcus aureus, and Haemophilus influenza.

1.4.2 The Flu Battle Ceases

The First World War was the first war involving hundreds of millions of people, with the flames of war spreading globally. Positional warfare, tanks, and new fighter jets claimed countless lives. However, the end of World War I was flu-related.

On 4 March 1918, the first flu case at Camp Funston in the State of Kansas, United States began to develop symptoms. By noon, the number of patients exceeded 100. Within three weeks, 1,100 soldiers were hospitalized as their condition worsened. As the troops continued to move between military camps, the flu quickly spread within the camps.

In 1918, World War I was ongoing, and the war in Europe was heating up. The US government decided to exercise strict control of the news while continuing to send troops to Europe. In March 1918, 84,000 American soldiers marched towards the European front. One month later, another 118,000 American soldiers crossed the oceans to engage in the war.

In early April, influenza began appearing in Brest, where the Americans landed. The French naval headquarters in Brest were severely impacted by this. The flu continued marking its way across the European continent and the rest of the world at an alarming rate, causing an unprecedented disaster.

This attack can be categorized into three waves. The first wave, which was between April and July 1918, spread rapidly from Brest to the whole of Europe. It was characterized by high morbidity and low mortality. The second wave, which was from July to November of the same year, swept across Europe, America, Asia, and the African continents. It was characterized by very high morbidity and mortality, with the 20–35 age group, who were the main forces of the war facing the highest mortality rate. The third wave began during winter 1919 and gradually mysteriously disappeared by the spring of 1920. In Australia, the flu lasted until August 1919, which was winter in the southern hemisphere. In Hawaii, it lasted until March 1920.

The flu, which shocked the world, was the Spanish flu. However, this flu was called the "Spanish flu" because it originated from Spain. At that time, all countries at war censored the news, and the flu pandemic was not made known to the outside world. As Spain was not part of the war, it did not censor news on the pandemic and its media carried more reports on it.

During the first two waves of attacks, 548,000 people died in the United States, accounting for 0.5% of the national population. In 1918, the average life expectancy of the American population decreased by 12 years. The United States had to mobilize every force to fight the flu, and the mobilization of its population to support the war became a mobilization to fight the flu.

In the United Kingdom, 215,000 people perished. In April 1918 alone, 31,000 British troops were infected. By May, 10% of the Royal Navy was infected and unable to fight for three weeks. The average weekly death toll in England hit 4,482, and even King George V was infected.

The death toll in France was 166,000. On average, 1,200 people died weekly in Paris. In early June, when the German army launched a fierce attack, nearly 2,000 French soldiers had to withdraw from the battlefield because of the flu.

Its Allies' situation was equally bad.

The death toll in Germany was 225,000. In March, to conclude the war before the arrival of large American troops, the Germans launched a series of offensives. The brave German soldiers rushed into the trenches of their opponents and received the influenza virus left behind by them. Young people, which experienced the highest mortality rate, were the main force supporting the war. The Governments were eager to conclude the war and concentrate on fighting against the more horrifying enemy.

In November 1918, the revolt of the German navy in Kiel, Germany, triggered a general revolution, which spread all over Germany. Its war machines were the first to cease fire. After Kaiser Wilhelm II abdicated, the German government appealed to the Allied Powers, which no longer had the resources to fight.

On 11 November 1918, German government representative Erzberger and Allied Commander-in-Chief Foch signed an armistice agreement at Le Francport in the Forest of Compiègne in northeastern France. The victorious Allied countries fired a 101-gun salute, declaring the end of World War I.

1.5 History of Ebola

The common harmful viruses, which we encountered, come from a huge viral family. They can be divided into the spherical-shaped virus, baculovirus virus, brick-shaped, coronavirus, chain virus, and filovirus according to their shapes. When the filovirus is mentioned, most non-communicable disease or virus experts may not be familiar with it, but most will know the Ebola virus, the star of the filovirus family.

The Filoviridae family currently contains three genera, namely Ebolavirus, Marburg virus, and Cuevavirus. Among them, the genus Ebolavirus is further organized into five species, namely Zaire, Sudan, Tai Forest, Bundibugyo, and Reston.

1.5.1 Origins of Ebola

The discovery and confirmation of the Ebola virus can be traced to 1976 in Ensala, South Sudan (which was part of Sudan at that time). On 27 June 1976, the first known case in Sudan involved a cotton factory worker in Nzara, who was hospitalized on June 30 and died on July 6. Although the WHO medical personnel involved in the Sudan pandemic knew that they were dealing with an unknown disease, it was not until a few months later that they got a "positive identification" and named the virus. The Sudan outbreak infected 284 people and caused 151 deaths.

On 26 August 1976, a second Ebola virus outbreak occurred in Yambuku, a small village in the northern Mongala district of northern Zaire (now the Democratic Republic of the Congo). This outbreak was caused by the Zaire ebolavirus, a different type of species from the first outbreak in Sudan.

The headmaster of Mabalo Lokela, a village school, was the first person to contract the disease. He began displaying symptoms on 26 August 1976 after visiting the Ebola River from 12 to 22 August in northern Zaire near the border of the Central African Republic. Initially, he thought he had malaria and took quinine, but he did not get better after taking it. His condition continued deteriorating and he passed away on 8 September, 14 days after the onset of symptoms.

Shortly after Lokela's death, those who were in contact with him showed similar symptoms and died, and the people in Yambuku began panicking. As the pandemic ensued, researchers' identified it as a new strain related to the Marburg virus. The strain samples separated from the two outbreaks were near the first confirmed outbreak location in Zaire. Hence, the new virus was named after the Ebola River, namely the "Ebola virus."

During the 1976 Ebola virus outbreak in Zaire, there was a total of 318 cases and 280 deaths, equivalent to an 88% mortality rate. These outbreaks were initially thought to be related. However, scientists later realized that they were caused by two different Ebola viruses, namely the Sudan and Zaire species. With the World Health Organization's assistance, the pandemic was brought under control. The villagers were isolated, while medical equipment was disinfected, and protective clothing was provided for transportation by the Congolese Air Force.

Throughout the 1980s, the Ebola virus remained silent in Africa. However, the experts did not expect it to re-emerge at another location, which was deemed unthinkable.

Reston, located in northern Virginia, near Dulles International Airport and 20 miles away from the District of Columbia, is one of the major IT towns in the United States. In 1989, the Reston Primate Quarantine Unit of Hazelton Research Products, a subsidiary of Corning Incorporated, received a batch of long-tailed macaques imported from the Philippines for medical experiments. Two were found dead on arrival. However, they did not raise any eyebrows as long-distance transportation of similar animals often led to some deaths.

However, deaths continued occurring at the Primate Quarantine Unit in Reston, and the death rate of two to three monkeys daily meant that it was possibly an infectious disease, which alarmed the vigilant center's staff. Further observations by the researchers revealed that the dead monkeys had symptoms of hemorrhagic fever, similar to Ebola virus infection.

This raised the U.S. Diseases Control and Prevention threat assessment to the highest level. After the United State Army Medical Research Institute of Infectious Diseases received the news, the military killed the remaining monkeys and disinfected the center, before demolishing and rebuilding it. The Institute later confirmed that the Ebola virus caused their deaths, but it was different from the Zaire and Sudan species. Six staff in the center were found to have antibodies to the virus in their blood, which indicated that were infected. As such, experts believed that the virus could have been airborne.

In other words, this species of the Ebola virus could spread the influenza virus. Therefore, concerning the situation and mortality rate in Zaire and Sudan, as well as the dense population in the suburbs of the US capital, a catastrophe was imminent. Apart from the relevant personnel being fearful of death, there was an uproar among the Washington community. The high-level officials were also perturbed by the developments. They were unaware of the virus spread, let alone of what would happen to the Washington area and the United States if the Ebola virus did strike.

Just as the thoughts of an impending storm began to haunt, the infected few did not get sick. During an investigation by experts in the Philippines, a large number of monkeys, which died of Ebola virus infection, were found at a monkey exporter near Manila. In addition, 12 Filipinos found to have developed antibodies to the virus in their blood were asymptomatic.

It was not until 2008 that this paradoxical situation was explained by the Institute. It was because it was related to another species known as Reston ebolavirus. At that time, while investigating a pig disease outbreak in the Philippines, researchers discovered that some animals were infected with equine arteritis virus (porcine reproductive and respiratory syndrome virus) and Reston ebolavirus.

Serology tests showed that while a small number of Philippine pig farmers had IgG antibodies against this virus species, they did not develop serious symptoms. This provided additional evidence that it could cause mild or asymptomatic infections in humans.

1.5.2 Ebola Across Borders

In 1976, the Ebola virus was first discovered, and its first outbreak resulted in 315 deaths. In 1995, its second major outbreak occurred in Zaire (now the Democratic Republic of Congo, DRC) and caused 254 deaths.

In 2000, the Sudan ebolavirus broke out in Uganda, infecting 425 people and causing 224 deaths. In 2003, an outbreak in the Democratic Republic of the Congo (DRC) infected 143 people and caused 128 deaths (90%). This was Ebola's highest fatality rate to date. In 2004, a Russian scientist died from the virus after being pricked with a contaminated needle.

Between April and August 2007, people with fever at four DRC villages were confirmed Ebola cases in September of the same year. It caused 264 infections and 187 deaths.

On 30 November 2007, after confirming samples tested by the National Reference Laboratory and the Centers for Disease Control, the WHO confirmed the existence of an Ebolavirus species in the Bundibugyo District in western Uganda and named it the Bundibugyo type. WHO reported 149 cases, including 37 deaths.

On 17 August 2012, the DRC Ministry of Health reported an outbreak of the Bundibugyo-type in the eastern region. According to the WHO report, it resulted in 57 infections and 29 deaths and was probably caused by bush meat hunted by local villagers around the towns of Isiro and Viadana.

In March 2014, the Ebola pandemic reemerged in Congo (Brazzaville). However,

contrary to previous pandemics, many highways were open to traffic after 2011, resulting in an increased movement of people among the three countries of Congo (Kinshasa), Congo (Brazzaville), Sudan, and other African nations. Hence, the virus made its way to West African nations such as Guinea and Liberia.

After making its way through Equatorial Africa, it attacked the area like a tiger among a flock of sheep. Medical personnel in West African nations such as Guinea and Liberia were infected with the Ebola virus due to their direct contact with patients. During its incubation period of 21 days, the hospitals became the primary source for virus transmission. In August, new infections in Liberia doubled every 15–20 days. On 8 August, the WHO acknowledged that it underestimated Ebola's spread and potential and declared it an "international public health emergency."

As of 2014, the Ebola virus has been raging in West Africa for nearly 30 years, but our understanding of it remains limited. There are no vaccines or approved treatments thus far. The only option for doctors is to give patients fluids to let them stay hydrated until their immune systems can defeat the virus.

On 19 September 2014, the decisive turning point occurred when the United Nations established an Ebola emergency response mission. More than 20 countries deployed medical personnel and supplies to assist in the fight against Ebola. On 29 December 2015, 42 days after the last person tested negative twice, Guinea was declared free of Ebola virus transmission. In 2016, the WHO confirmed the end of the epidemic.

In this epidemic, there were more than 28,000 cases, including over 11,000 deaths. There were also cases in the United States, Britain, France, and Italy, which caused a global panic.

1.5.3 How Terrifying is Ebola?

Since the Ebola virus was discovered, academia has been conducting detailed research on the Ebola virus.

It is generally believed that the tradition of consuming wildlife and unique funeral rituals are the important reasons behind the continuation of Ebola hemorrhagic fever in four West African countries. Since its discovery in 1976, Ebola hemorrhagic fever outbreaks have occurred more than 20 times, but they have been concentrated in the forested areas of equatorial African countries such as Sudan, Congo (Brazzaville), Congo (DRC), Uganda, and Gabon. This is related to the unique practices of wildlife consumption and funeral rituals.

The local tradition of eating "bush meat" in West Africa has continued for thousands of years, including fruit bats, host animals of the Ebola virus, as well as indirect primates hosts such as gorillas. Although the West African countries have already banned the consumption of wild animals such as bats and monkeys, "bush meat" remains a protein source that people rely on for survival in West Africa, where famine is prevalent. In addition, there is no local tradition of cremation. Residents touch and kiss the remains as they bid farewell to the dead. Sometimes, the whole village participates in traditional funerals, and they could become one of the main places for its spread.

The Ebola virus is one of the deadliest and most terrifying diseases known to mankind. Initially, the patient will have a fever, headache, and a sore throat, followed by abdominal pain, vomiting, and diarrhea. As the condition worsens, many patients will become unresponsive, sluggish, and develop purple rash and hiccups.

The scariest symptoms occur a few days after the illness. Ebola virus-infected cells invade the inside of blood vessels, causing bloody fluids to leak from the mouth, nose, anus, vagina, and even eyes. The virus damages the liver badly, destroying cells and affecting its ability to produce clotting proteins and other important components in plasma. Eventually, the patient's blood pressure drops significantly, leading to shock and multiple organ failure, making it impossible to reverse the situation.

While Ebola hemorrhagic fever is less infectious, it is highly contagious. About half of Ebola virus-infected patients experience bleeding symptoms. Human-to-human transmission takes place through contact with body fluids. The R0 value of the Ebola virus (Africa) is about 1.5–2.5, which is less infectious as compared to other respiratory diseases such as the SARS virus and coronavirus. However, Ebola is highly infectious. 1cm3 of blood contains 1 billion viruses, and the probability of infection by contact with a patient's body fluids without protection is almost 100%.

In terms of transmission, it is generally believed that the Ebola virus outbreak began after individuals came into contact with tissues or body fluids of infected animals. Once a patient becomes sick or dead, the virus will continue its spread to individuals in direct contact with an infected person's blood, skin, or other bodily fluids. Studies involving primate laboratory animals found that inoculating virus-containing liquid droplets into their mouth or eyes can cause them to be infected with the Ebola virus, which suggests that human infection can take place through an infected hand to these locations.

For symptomatic patients with exposure history, medical evaluation generally includes testing for the Ebola virus and suspected pathogens. Whether a laboratory test is required for the Ebola virus depends on the relative likelihood of the patient's exposure to the virus, and if the clinical symptoms and/or laboratory test results match it. The diagnostic test for Ebola virus infection is mainly completed to detect specific RNA sequences in blood or other body fluids by RT-PCR. While the virus can be detected in blood samples within three days following symptoms onset, a retest may be required if the sample was taken less than three days apart.

Regarding the treatment of Ebola virus disease, although there is no fully effective vaccine and medication, people have been able to improve its treatment. In 2019, four drugs and two vaccine products were approved for emergency use during the Ebola pandemic. However, they are under Clinical Phase III, with their effectiveness pending verification.

The four main drugs include three monoclonal antibodies and one RNA polymerase inhibitor, comprising multiple monoclonal antibodies ZMapp (Mapp) and REGN-EB3 (Regeneron Pharmaceuticals), as well as single monoclonal antibodies mAb114 (Ridgeback) and Remdesivir (Gilead Sciences). Among the four drugs, Mabel14 and REGN-EB3 are the most effective as they result in the highest reduction of viral concentration. The test results came back negative after 15 – 16 days of taking these drugs, while the time for test results to turn negative after taking ZMapp was about 48 days. Compared with the other three drugs, the effect of Remdesivir was relatively limited.

Only two Ebola vaccines were approved. One of them is rVSV-ZEBOV-GP (ERVEBO), a live, attenuated vector vaccine by Merck & Co and certified by the European Union and US FDA on 12 November 2019. It is solely used for active immunization of people aged 18 and above to mainly protect against the Zaire-type Ebola virus. The second is the recombinant Ebola vaccine (adenovirus vector), developed by the Academy of Military Medical Sciences and approved by the China Food and Drug Administration on 19 October 2017.

1.6 "Self-healing" AIDS

Since the first AIDS report by the US Centers for Disease Control and Prevention on 5 June 1981, the virus has spread globally over the next 40 years, with more than 36 million people infected with AIDS before 2020.

According to a joint assessment by the Center for Disease Control and Prevention, as well as the Joint United Nations Programme on AIDS/HIV and WHO, as of end-2018, China estimates that about 1.25 million AIDS-infected people are living in China, with about 80,000 new infections annually.

In terms of gender, among the newly discovered HIV infections in China annually, the number of men is about three times that of women. In 2017, in terms of age, nearly 800 newly discovered HIV-infected persons were below the age of 15. In terms of transmission routes, China's AIDS prevention and control efforts achieved remarkable results. Transfusion-transmission methods were eradicated, alongside the effective control of transmission through injection drug use. The mother-to-child transmission rate is also at the lowest level in history. Currently, sexual transmission is the primary way of AIDS transmission, accounting for over 95%.

1.6.1 Four Stages of AIDS

The medical term for the abbreviation AIDS is "Acquired Immune Deficiency Syndrome." It is a serious disease involving the human immunodeficiency virus (HIV) attacking and destroying the human immune system before causing their eventual death.

The Acquired Immune Deficiency Syndrome comprises of three aspects: it is acquired, it has immunodeficiency, and it is a syndrome. Acquired means the condition is acquired, meaning that a person becomes infected with it. In terms of pathogenesis, immunodeficiency means that it is primarily caused by human immune system damage, leading to a reduction and loss of its defense mechanisms. The common characteristics of immunodeficiency diseases include the (1) increased susceptibility to infection, (2) high likelihood of malignant tumors occurring, and (3) diversified clinical manifestations. In the context of clinical symptoms, syndrome refers to a complex group of symptoms, such as opportunistic infections and tumors in various systems caused by immunodeficiency.

As the AIDS-causing HIV can attack CD4 T cells, the command center of the immune system, paralyze it and cause extremely high mortality. As such, patients often need to receive antiretroviral therapy (ART) to manage its progression. More importantly, as an RNA virus, HIV has a high genomic mutation rate, making it very difficult for AIDS vaccines and medication to be developed. In February 2020, the clinical trial of HVTN 702, a highly regarded AIDS vaccine candidate in the scientific world, failed. Hence, its high mutation and drug resistance rates made HIV an "incurable disease."

HIV, which causes AIDS is mainly present in the blood, semen, vaginal secretions, breast milk, and body fluids of infected persons and patients. This is closely related to the three main AIDS routes, which are sexual, mother-to-child transmission, and blood.

Sexual transmission: Unprotected sex with HIV-infected partners, who are homosexuals, heterosexuals, and bisexuals. Mother-to-child transmission: HIV-infected mothers may transmit it to the fetus and infant via pregnancy, childbirth, and breastfeeding. Blood transmission: Intravenous drug use and sharing of unsterilized injection tools used by HIV-infected with others are key HIV transmission routes. These include blood transfusions, blood products, medical equipment, such as dental equipment, surgical equipment, and shared razors, toothbrushes, tattoos, eyebrow tattoos, and beauty equipment contaminated with HIV in our daily lives.

The four clinical stages of AIDS range from the acute, asymptomatic, pre-AIDS to AIDS stages.

Acute stage: People infected with AIDS will develop acute symptoms of HIV infection about 10–14 days after engaging in high-risk behavior and this lasts 1 to 2 weeks. At this juncture, they are called "HIV carriers." They may present typical symptoms such as unexplained fever, cough, abdominal pain, diarrhea, fatigue, nausea, vomiting, rash, superficial lymph nodes, and sore throat, etc. It is worth noting that more than 50% of HIV-infected people do not experience acute symptoms at this stage, hence they should not be relied upon for assessment.

Asymptomatic stage: After the acute stage, the immune system suppresses viral activity and can reduce its load in the blood. The patient then enters the clinical incubation period of HIV. Its length is affected by many factors and ranges from two weeks to 20 years. The patient does not present noticeable symptoms during this period.

Pre-AIDS stage: After the asymptomatic infection period, noticeable signs and symptoms associated with AIDS which appear are called AIDS-related complex or persistent generalized lymphadenopathy. They include persistent lymph nodes enlargement, beginning from the neck, followed by axillary and inguinal lymph nodes, enlargement of no more than two lymph nodes, weight loss of above 10%, periodic fever (around 38 degrees Celsius) lasting for several months, night sweats and infections such as herpes simplex virus and Candida albicans (a type of fungus).

AIDS stage: As the immune system is severely damaged, various fatal opportunistic infections and tumors are highly likely to occur. Lesions can appear in the lungs, mouth, digestive system, nervous system, endocrine system, heart, kidneys, eyes, joints, and skin, etc. The average survival rate of those with an opportunistic infection is nine months.

In reality, it is not possible to detect the infection immediately after infection, because there is a window period, which is the time between an HIV infection and detection of antibodies or nucleic acids in the blood.

At present, this period differs, depending on the kind of test taken. In general, the window period is about one week for nucleic acid tests, about a fortnight for fourth-generation antigen-antibody tests, and about three weeks for third-generation antibody tests. By this time, most people living with HIV can be identified.

When an infection cannot be confirmed, antiretroviral drugs, known as Post-exposure Prophylaxis, can be used to prevent the spread of HIV.

The principle of post-exposure prophylaxis (PEP) is to stop HIV from replicating and prevent it from spreading from infected cells to more cells. Using sexual transmission as an example, the virus first invades the mucosa and passes through its barrier, before entering the body's tissues, cells, and lymph nodes and multiplying within and entering the blood thereafter. Its purpose is to kill the virus before it reaches the blood.

After exposure, the sooner PEP drugs are taken, the quicker the blood concentration of the drug can increase to ensure that it takes effect before the virus enters the blood. This is a process involving the drug's race against the virus. The golden period is two hours with a success rate of over 99%. This rate will gradually decline thereafter, but there is still a high success rate within 72 hours, also known as the golden 72 hours.

PEP drugs are highly effective, but not many people know about them. According to China Center for Disease Control and Prevention statistics in 2017, nearly half of the 10,000 netizens interviewed did not know that PEP drugs can prevent AIDS.

1.6.2 The Elite Controllers Who Do Not Take Medication

Under normal circumstances, a virus will be violently attacked by the immune system when it enters the human body. White blood cells will respond immediately by releasing antiviral proteins, attacking infected cells while seeking additional support. However, HIV can break through the immune system limits and infect helper T cells, which are critical immune cells that help the body fight pathogens.

HIV will first adsorb and penetrate the cell and will bring its enzymes and genetic material close to the nucleus once it is inside. Reverse transcriptase will convert viral RNA into proviral DNA, which will be inserted into the cell's genome, and force the cell to spit out HIV protein and genetic material to replicate a new copy of the virus.

HIV inserts its genetic information into the human genome of an infected person known as the "provirus." They exist in the human body for a long time, awaiting an appropriate time to be "activated," before replicating in large numbers again, which is why HIV is difficult to eradicate. Therefore, HIV-infected people need to take ART drugs to inhibit virus replication.

However, there are always exceptions. Approximately 0.5% of HIV-infected people seem to have an innate ability to control the virus and can inhibit the virus from replicating in the body without receiving ART treatment. These people are called "elite controllers."

HIV-infected persons known as "elite controllers" do not need to take ART drugs to keep their viral load at undetectable levels for years. Likewise, a study published in the journal *Nature* provided the answer. Research led by Yu Xu, a Chinese scholar and Associate Professor at Harvard Medical School, found that among these elite controllers, viruses often integrate into specific regions of the human genome, where transcription is inhibited.

The research team used the latest sequencing technology to accurately map the position of HIV genomes in the human genome, and compared the provirus in the cells of 64 people, who maintained HIV-1 drug-free control, with 41 people, who were receiving ART. The three main findings are as follow:

Firstly, while elite controllers have lesser proviruses in their bodies, they have a greater proportion of full proviral sequences. There are four elite controllers with full proviral sequences accounting for 100% of detected species, meaning that they have the potential to produce infectious virus particles during transcription. It is worth noting that there is limited evidence of escape mutation, which shows that they were produced in the early phase and exist for a long time.

Secondly, among the "elite controllers," 45% of the proviruses were located in the "gene desert," which are chromosomal regions where transcription rarely occurs. Among people receiving antiretroviral therapy, the percentage is 17%.

The researchers used matched integration sites and proviral sequence (MIP-seq) analysis to study the sites where the virus has integrated into the host genome. It was found that a complete proviral sequence in the elite controller is more likely to be integrated into the non-protein-coding region of DNA or KRAB-ZNF gene on chromosome 19. These regions are composed of heterochromatin with low transcriptional activity and are usually not conducive for HIV-1 integration.

This means that HIV has been locked in the cell's genome and was prevented from replicating into more viruses, thus preventing an infection.

Thirdly, the researchers analyzed the accessible chromatin regions (regions that may be transcribed) and found that viral integration sites of elite controllers' DNA are often farther away from the host genome where transcription begins. This also shows that the controllers' genome is unlikely to actively produce viral transcripts and proteins.

The study of elite controllers proved the possibility of functional therapy (HIV), at least in principle. It showed that HIV-1 infection can be cured spontaneously through natural immune-mediated mechanisms.

1.6.3 "Self-Healing" AIDS Patients

In the December 2010 issue of American magazine *Blood*, researchers from medical departments from three German universities published a paper with strong findings that a 43-year-old male AIDS patient was cured. Thereafter, on 26 August 2020, a leading academic journal *Nature* published a major paper on AIDS online. The research was led by Professor Xu Yu, an immunologist from Lagan Institute, United States of America. The study revealed how a group of HIV-infected individuals could spontaneously control viral replication in their bodies without antiviral drugs. Among them, there was a rare case of a drug-free controller, who was seemingly cured of HIV. This discovery may rewrite the history of mankind's fight against AIDS.

The paper mentioned that the untreated HIV-1 patients who were able to control HIV-1 replication below the limit of detection for a prolonged period were usually called "elite controllers." This may be the closest to a natural cure for HIV-1 infection. Previous research by scientists has linked the above to specific mutations in HLA class I genes and the existence of highly functional cellular immune responses. Academia has been researching elite controllers for decades. After the world's first "spontaneous recovery" of AIDS patients was reported, it has once again garnered the medical community's close attention on HIV self-healing. As of 2020, there have been four cases of HIV DNA disappearance, which are considered to be close to the "curve" of HIV. Three of these cases were published in important journals, while the other case was reported at AIDS 2020. Their details are as follows:

In 2009, *ENJM magazine* reported the first HIV self-healing case, also known as the "Berlin patient." In 2007, Timothy Ray Brown, who was suffering from AIDS and leukemia, received a bone marrow stem cell transplantation in Berlin, Germany. The said bone marrow donor had a rare genetic mutation (CCR5), which made immune cells HIV-resistant.

One year later, Timothy Ray Brown received another bone marrow transplant from the same donor due to leukemia relapse. He became homozygous and experienced a strong graft versus host rejection (GvHD) after the transplant, and both diseases disappeared from his body thereafter. He stopped antiviral treatment one day before the first bone marrow transplantation to 548 days after the first bone marrow transplantation. He has not relapsed thus far.

A decade later, a second HIV self-healing case was reported globally. On 5 March 2019, the journal *Nature* published a paper titled "*HIV-1 remission following CCR5Δ32/Δ32 hematopoietic stem-cell transplantation*" introducing the "London Patient" as the second case of successful stem cell transplantation for AIDS for the first time.

The said HIV patient, who received a CCR5delta32 bone marrow transplantation due to Hodgkin's lymphoma, quickly became a homozygous donor and developed mild GvHD. He stopped taking antiviral drugs 16 months after receiving a bone marrow transplant and did not experience a viral rebound for the next 18 months.

Its viral load in peripheral blood was less than 1 copy/mL (below the detection limit), reflecting that the VOA method, the gold standard detection method for the virus reservoir size, cannot detect any complete viral reservoir in the patient. More importantly, the said patient's HIV antibody titer gradually decreased, and the blood serum eventually produced a negative result. This was the second case after the Berlin patient.

In 2020, Nature magazine reported another HIV self-healing case, known as the San Francisco patient. This case, an extreme case of an elite controller, was the only case that healed without any treatment administered. The patient did not take medication after being diagnosed with HIV in 1992. The patient underwent 34 viral load tests during the 24 years from 1995 to 2019, and the viral load was always below the detection limit.

Although there was the integration of HIV in its genome, there was no complete HIV full-length DNA in 1.5 billion PBMCs and no complete HIV full-length DNA in 14 million resting CD4. The qVOA testing method involving 340 million resting CD4 did not replicate the competent HIV virus. This special case of an elite controller under exceptional circumstances showed that through the viral-host, HIV infection can also undergo a self-limiting elimination, which meant that HIV could heal on its own.

The last "Sao Paulo patient" was reported at the 23rd International AIDS Conference (AIDS 2020). This case came from the SPARC-7 clinical trial conducted in Brazil in 2014, where 30 subjects were given DTG, CCR5 Antagonist Maraviroc, and a depot activator nicotinamide (NAM).

Among them, a 35-year-old patient was diagnosed with HIV in October 2012, underwent AZT/3TC + EFV antiviral therapy in December 2012, and joined the SPARC-7 study as mentioned above in September 2015. The study ended in September 2016, and the patient stopped ATI in March 2019. During the trial, the patient experienced a brief blip, which proved that the reservoir may be released by NAM.

An ATI viral load, which does not rebound, is the gold standard for evaluating HIV cure. The patient's viral load was undetectable during the ATI period. Amazingly, before ATI and after NAM treatment, the said patient's HIV DNA gradually became undetectable. The procedure, VOA assay was unable to release the HIV virus, which reflected that its reservoir was gradually disappearing. What is even more amazing is that during the ATI stage, the HIV antibody turned gradually negative, alongside the decreasing activation of CD8 cells.

2

Memories of SARS

2.1 How much do you know about SARS?

Coronavirus is named after its iconic coronal structure, a radial spike formed by S glycoprotein on the viral surface, seen under the electron microscope. The microstructure shows that it contains two main envelope proteins, namely S glycoprotein and M protein. The former is the main antigen for receptor binding and cell fusion, while the latter is involved in the budding and enveloping formation process and it plays a key role in the assembly of enveloped viruses. A few coronaviruses also have a third glycoprotein – hemagglutinin esterase.

Before the SARS outbreak, no one knew that it was so highly infectious.

Humans are now aware of its existence. Like a virus that can infect the respiratory and digestive tracts of animals and humans, it has not been receiving enough attention from us for too long. This ignorance is justifiable for healthy people with normal immunity as it only causes very mild symptoms.

However, SARS struck in the first decade of the 21st century. According to WHO data, as of July 2003, the SARS virus had caused 8,096 infections and 774 deaths in 27 countries. Ten years later, the Middle East Respiratory Syndrome caused by the MERS virus resulted in 1728 confirmed cases across 27 countries and claimed 624 lives. This century has just entered its third decade, and a novel coronavirus has returned. It was also during this process that people gradually began facing it and studying it.

Studies have found that its high pathogenicity is dependent on its flexible gene recombination and rapid adaptive mutation. On one hand, the RdRP (RNA-dependent RNA polymerase) used in replicating single-stranded RNA has an inherent 1,000,000 mutation/site/replication error rate, which can lead to continuous point mutations. On the other hand, when two types of coronaviruses infect the same host, hundreds or thousands of base pairs of genome fragments can be obtained from the genome of the other virus type to increase their ecological locus or become a new virus. This has caused the coronavirus to rapidly mutate into three new human coronaviruses, namely SARS, MERS-CoV, and 2019-nCoV, with great pandemic potential in approximately twenty years.

Coronaviruses can be divided into four categories according to their genomes and structures. Among them, α and β infect mammals only, while γ and δ primarily infect birds. The first known global coronavirus was the avian infectious bronchitis isolated in 1937 and confirmed to be the pathogen causing severe infection in chickens. The first human coronavirus was isolated from the human nasal cavity in 1965. Its in vitro amplification results showed that it has been existing in humans for at least 500–800 years and had originated in bats.

For a long time coronavirus was an important animal pathogen, and it could cause respiratory and intestinal diseases in mammals and birds, as well as mild and self-limiting upper respiratory tract infections in humans, such as the common cold (15–30%), pneumonia and human gastroenteritis. Six known coronaviruses can cause human diseases, including HCoV-229E, HCoV-OC43, HCoV-NL63, HCoV-HKU1, SARS, and MERS-CoV. The first four are epidemic diseases, which mainly cause mild self-limiting diseases, while the remaining two can result in severe illnesses.

SARS and MERS-CoV, discovered in 2002 and 2012 respectively, are β-coronaviruses and placed under the WHO High Threat List due to their huge threat to humans. Their high infection rates pose a continuous threat to human health. The 2019-nCoV, isolated from a large-scale respiratory disease outbreak in Wuhan, China at the end of 2019, was characterized as a β-coronavirus and became the seventh discrete coronavirus species capable of causing human diseases.

2.2 History of SARS

When speaking of "atypical pneumonia," many people may recall it as "SARS." In fact, SARS was known as "atypical pneumonia" initially, but this was done just to distinguish it from typical pneumonia. Later, SARS was named "Acute Severe Respiratory Syndrome," but it was called SARS as the 2003 outbreak left a deep impression on so many.

2.2.1 Tracing SARS

On 24 March 2003, the pathogen of SARS was first reported by Hong Kong, China, and the Centre for Disease Control and Prevention in the United States and supported by various evidence such as cell culture, microscopy, microarray data, serological testing, and PCR. The WHO officially announced that it was a newly discovered member of the coronavirus family, named "Urbani SARS-associated coronavirus" (abbreviated as SARS) on 17 April 2003. Its sequencing results showed its close relations to known characterized coronaviruses and it was not found in humans.

During the virus tracing process, no SARS antibodies were found in healthy people, indicating that the virus had not been spread among people. It was likely originated in animals and caused human infection after mutating. Researchers in the United States and the Netherlands established a monkey infection model system and finally proved that it was a pathogen.

Early studies showed that SARS was related to bovine and murine hepatitis coronaviruses. However, no bovine-mouse source was detected in its sequencing results. Therefore, the researchers inferred that it was a new unknown pathogen and it did not originate from an existing virus strain. Subsequent studies predicted that it may have derived from the predecessors of coronavirus as wild animals were naturally infected before it crossed over the barrier of human species to cause SARS and then produced by recombination of animal and human viruses.

In May 2003, Hong Kong scientists reported that a virus, similar to the virus that caused SARS, was found in a rare civet cat (a type of arboreal cat). Hong Kong researchers said that the coronavirus was found in the feces of palm civets living in Pakistan-Indonesia. In November 2002, operators of processed meat and chefs from Guangdong Province, who participated in banquets, were among the earliest SARS cases. The research team was able to cultivate a coronavirus almost identical to the SARS virus from 25 civet cats, representing eight different tested species.

Another research team detected SARS-like viruses in live animals sold in Guangdong Province. Concurrently, studies found that the virus was found in Himalayan palm civets (Paguma larvae) and raccoon dogs, while sequencing results showed phylogenetic differences of the virus between animals and humans. Subsequent studies showed that its infectious period in civet cats was not long, and other species, such as bats, may have also been its natural hosts.

The relevant personnel conducted a serological analysis of the affected area, and results showed that about 40% of wildlife dealers and 20% of slaughterhouse workers carried SARS antibodies. The researchers speculated that SARS-like coronaviruses existed in the area at least two years before the SARS outbreak. While it was not infectious among the population initially, it became SARS after an evolutionary adaptation.

To date, we are still unable to ascertain its origin. On 1 November 2013, the *Science and Technology Daily* reported that Shi Zhengli's research team at the Wuhan Institute of Virology,

Chinese Academy of Sciences isolated a SARS-like coronavirus (SARS-like CoV) highly homologous to SARS, further confirming that the Chinese rufous horseshoe bat is the source. These results were published online in the journal *Nature*. To date, this remains a rather authoritative conclusion.

2.2.2 The Pain of SARS

From the perspective of virus transmission, SARS can be spread through droplets and direct contact. Its concentration can reach about 100 million particles/ml and it survives up to six days on contaminated surfaces and objects at room temperature.

In addition, SARS primarily attacks the respiratory tract, in particular the lower respiratory tract, causing severe acute viral pneumonia. WHO's definition of suspected SARS cases includes high fever (>38°C) or a history of fever in the past 48 hours, chest X-ray examination revealing new infiltration of pneumonia, flu-like symptoms (chills, cough, malaise, and myalgia), or history of exposure to SARS, and one or more positive test results for SARS. However, the initial symptoms and clinical manifestations of SARS are difficult to distinguish from other common respiratory infections, and the elderly population may not experience fevers.

Analysis of autopsy samples and experimentally infected animals showed that SARS will affect the lungs and can be detected in type 2 lung cells. Studies have indicated that SARS often causes diffuse alveolar damage, bronchial epithelial peeling, loss of cilia, and squamous metaplasia in tissues. In some cases, giant cell infiltration, hemophagocytosis, and enlarged alveolar epithelial cells were also observed. The infection passes through an inflammatory phase or an exudative phase (characterized by hyaline membrane formation, lung cell hyperplasia, and edema), a proliferative phase, and a fibrotic phase. Apart from respiratory infections, there were gastrointestinal and central nervous system infections.

Clinical characteristics revealed that the display of SARS symptoms usually follows a three-stage pattern. In the first week after infection, the usual symptoms include fever and myalgia and they may be related to the direct viral cytopathic effect. The increase in viral load can be detected by PCR. During the second week, seroconversion is detected, followed by a decrease in viral load. Twenty percent of infected patients are clinically characterized by a deterioration of their condition and uncontrolled virus replication. This stage may be due to an excessive immune response triggering an immune pathological damage, a result of lung damage caused by SARS.

In terms of patient prognosis, a prospective study reported in 2007 provided a comprehensive long-term prognosis of SARS survivors for the first time. It studied 117 SARS survivors from Toronto, Ontario who underwent physical examinations, lung function tests, chest X-rays, and a 6-minute walk test at three different periods (3, 6, and 12 months) after discharge.

The results showed that most SARS survivors had fully recovered from the disease a year after infection. However, their overall health, vitality, and social function one year on were lower than normal levels, and many reported that they could not return to their pre-SARS levels at work. During the evaluation period, the utilization rate of mental health resources was significantly higher than usual. A subsequent study of 22 long-term survivors in Toronto determined that symptoms that lasted up to 20 months after the onset of illness include musculoskeletal pain, fatigue, depression, and sleep disturbances.

A long-term follow-up study by Hong Kong researchers also found that in the fourth year, 233 SARS survivors suffered from significant mental illness and persistent fatigue, while another Hong Kong study showed that medical workers suffered long-term damage.

Apart from public health, the SARS pandemic had a grave impact on the economy. The global economic loss was estimated at USD30 billion, while East Asia's annual growth rate dropped by one percentage point. The total economic impact on China was USD25.3 billion, with Beijing's tourism sector alone reaching 1.4 billion U.S. dollars.

2.3 MERS, The Sniper

On 13 June 2012, a 60-year-old man in Jeddah, Saudi Arabia, was admitted to the hospital with fever, cough, and shortness of breath. He was having a fever for seven days when he was admitted to the hospital. He died of progressive respiratory and renal failure 11 days later. A new type of coronavirus was discovered after a pathogenic test.

A thorough investigation found that the disease appeared in Jordan as early as March 2012. As of May 2013, it was renamed "Middle East Respiratory Syndrome Coronavirus" (MERS-CoV) by WHO because all cases were either directly or indirectly related to the Arabian Peninsula.

2.3.1 MERS Outbreak

Many cases were reported in multiple areas after MERS was discovered. It continued ravaging in Saudi Arabia in subsequent years, until an outbreak broke out in 2014, with 649 cases, and it continued spreading. As of 10 May 2015, Saudi Arabia confirmed 976 MERS cases, including 376 deaths. There was a total of 1090 cases in the Middle East, with confirmed cases accounting for more than 95% of the total MERS cases in the world, making it the world's epicenter for MERS.

According to a WHO report, the MERS epidemic in Saudi Arabia only eased significantly in 2016–2018. In 2016 and 2017, the number of reported cases was 249 and 234 respectively, reflecting a continuous declining trend. It showed a large decrease from 492 cases in 2015. In 2018, the number of reported MERS cases continued decreasing, with only 145 cases, which indicated that MERS was better controlled.

However, in early 2019, another MERS outbreak broke out. From January to April, the number of confirmed cases reported were 17, 69, 39, and 20 respectively, with the cases increasing at a decreasing rate in the second half of the year. Saudi Arabia confirmed a total of 203 MERS cases in 2019, an upward trend compared with 2018, while its spread continued. The MERS pandemic in Saudi Arabia is ongoing and continues to require close monitoring.

Since the MERS outbreak occurred, 27 countries reported MERS-CoV cases, including Italy, Netherlands, France, Germany, Italy, Tunisia, Malaysia, United Kingdom, United States, Iran, Egypt, Lebanon, and Turkey. The initial cases also ranged from two infections confirmed in the Middle East and the United Kingdom's cluster infections, including a hospital infection in Saudi Arabia, thus confirming it can spread from person to person during close contact. As of November 2019, countries have notified the WHO of 2494 human infections confirmed by laboratory tests and 780 related deaths (fatality rate of 37.1%).

However, statistical studies indicated that published epidemiological data only reflected the number of patients with clinical manifestations of MERS, but the incidence of asymptomatic cases are higher. According to a serum survey of individuals admitted by medical professionals and who participated in disease research from December 2012 to December 2013, about 45,000 people in Saudi Arabia tested positive for MERS-CoV serology. In addition, a study of travelers to countries affected by MERS from September 2012 to 2016 estimated that 3,300 were severely infected in these countries (Saudi Arabia, United Arab Emirates, Jordan, and Qatar), which is approximately two to three times the total number of confirmed cases.

On 20 May 2015, South Korea reported its first confirmed case, the largest MERS-CoV outbreak in a country or region outside the Kingdom of Saudi Arabia. The first confirmed patient traveled to four countries in the Middle East and returned to South Korea without symptoms. As of 11 September, WHO reported 185 laboratory-confirmed cases and 36 deaths in South Korea, as well as one case in China.

2.3.2 SARS or MERS

Middle East Respiratory Syndrome (MERS) is caused by Middle East Respiratory Syndrome Coronavirus, with most cases occurring in Saudi Arabia. The media often refer to it as the new SARS. However, in reality, although they are both coronaviruses, they have clear genetic differences and use different receptors when infecting the human body. MERS is relatively less infectious, but its fatality rate is higher than SARS.

MERS-CoV, a β-genus coronavirus, was first discovered in September 2012 in Saudi Arabia and named the "SARS-like virus" due to its similar clinical symptoms to SARS. It became the sixth known human coronavirus and it was considered related to respiratory infections in humans, pigs, cats, dogs, mice, and chickens.

Investigations in recent years showed that MERS-CoV is transmitted from camels to humans by a few strains. Infected camels may not show any signs of infection, but the virus can be transmitted through the nose, eyes, feces of infected animals, and even their milk and urine. It can also be present in their internal organs and multiply the quickest in deep lung tissues. However, the camel is the intermediate spreader (intermediate host), not the source or host of the virus. Studies showed that bats are the transmission hosts, but coronavirus is widespread in animals and apart from bats, rodents and wild birds can also be infected.

In comparison with the SARS outbreak in 2003, MERS is a relatively young disease. As such, not much is known about it. Since its first case was discovered in Saudi Arabia in September 2012, most confirmed cases reported in more than 20 countries were men aged 24 to 94, with an average age of 56 and concentrated in Middle Eastern countries. Cases reported in other countries were also directly or indirectly related to the Middle East as they had traveled, worked, and did business in the Middle East.

The incubation period of MERS is 2–14 days. The main clinical manifestations are fever, cough, shortness of breath, dyspnea, and other acute severe respiratory infection symptoms, with diarrhea, nausea and vomiting, abdominal pain, and other gastrointestinal manifestations being more common. Renal failure may occur in some cases, while a few cases are asymptomatic or present mild respiratory symptoms. In severe cases, lung failure and death can occur. Ninty-six percent of patients have previous underlying medical conditions, with diabetes, chronic kidney disease, chronic heart disease, and hypertension being the most common. The mortality rate after diagnosis is relatively high.

In terms of clinical and laboratory similarities, both MERS and SARS can cause acute respiratory distress syndrome. At present, there is no specific medication or vaccine for MERS, and only supportive treatment is available. Middle Eastern countries, especially Saudi Arabia, continue reporting MERS cases. Among them, most patients had contact with camels, drank camel's milk, or had contact with confirmed patients. Therefore, infection prevention and control measures are essential to prevent its spread.

2.4 Public Health Emergencies of International Concern

In January 2020, a new coronavirus broke out in Wuhan, referred to as COVID-19 (Corona Virus Disease 2019; COVID-19). Its pathogen was a novel coronavirus. It experienced three stages, namely local outbreak, community transmission, and large-scale transmission since it emerged in mid-December 2019.

Exposure to a seafood market ushered in a local outbreak mainly caused by people who came into contact with it. At that stage, most cases were related to an exposure history at the said market.

A community spread led to the pandemic stage. The virus spread in the community through people in contact with the seafood market and resulted in community transmission. Interpersonal transmission and cluster transmission occurred in multiple Wuhan communities and families.

The large-scale spreading stage was the result of a community spread. As it coincided with the Spring Festival, there was a mass movement of people. The pandemic expanded and it spread rapidly from Hubei Province to other parts of China.

On 30 January, the WHO declared the pandemic a Public Health Emergency of International Concern.

2.4.1 Possibilities of COVID-19 Transmission

The new coronavirus belongs to the genus β-coronavirus. Evolutionary analysis showed that it is most similar to the severe acute respiratory syndrome-related coronaviruses of the Chinese rufous horseshoe bat, with a nucleotide homology of 84%. Its nucleotide homology with the human SARS virus reached 78%, while its homology with MERS reached about 50%.

In terms of transmission routes, droplets and contact transmission through the respiratory tract are its main routes. Many places have detected the coronavirus in the feces of confirmed patients, confirming the risk of fecal-oral transmission of the virus. Other pathways such as aerosol transmission and mother-to-child transmission are awaiting confirmation by research.

Respiratory droplet transmission – this is the coronavirus's main mode of transmission. It spreads through droplets produced when patients cough, sneeze and talk and infects vulnerable groups when they inhale them.

Indirect contact transmission means that the coronavirus can also be transmitted through indirect contact with an infected person. Indirect contact transmission refers to droplets containing the virus, which are deposited on the surface of articles. These droplets will cause infection on the hands, which can become contaminated after touching these surfaces, and then mucous membranes such as the mouth, nose, and eyes. It was detected on the surface of door handles, mobile phones, and other items in the living environment of confirmed patients in Guangzhou and Shandong.

Fecal-oral transmission – this transmission route remains unclear. The coronavirus was found in the stools of recent confirmed first cases in Wuhan, Shenzhen, and even the United States. This showed that the virus can replicate and exist in the digestive tract and suggested the possibility of fecal-oral transmission. However, it is not certain that eating virus-contaminated food will cause infection and transmission.

Aerosol transmission: Aerosol transmission refers to the nucleus comprising protein and pathogens after the droplets lose moisture while they are suspended in the air. The nucleus can float for a distance in the form of an aerosol, resulting in long-distance transmission. Currently, there is no evidence that the virus is spread through aerosols. The WHO also believes that further evidence is needed to assess this possibility.

Mother-to-child transmission is a concern because it has been reported that a mother was a confirmed coronavirus patient who just had a child. A throat swab test performed on this child 30 hours after birth suggested that the child may have been infected through mother-to-child transmission. Of course, this requires more scientific research to confirm.

Human-to-environment transmission is the latest way for the new coronavirus to spread. This discovery was put forth by Zhong Nanshan, an academician of the Chinese Academy of Engineering. On 19 December 2020, he publicly proposed the new topic of its "environmental transmission" and said that there were recent sporadic local cases in some parts of China (Shanghai, Chengdu, Beijing, Dalian, etc.). The domestic epidemic prevention and control face two major risks, namely the influence of external factors, which is overseas imports, as well as new coronavirus spreading through the environment.

Based on the recent outbreaks in China, it was mainly due to two aspects, namely imported cases, and cold-chain food imports. The latter brought in the new coronavirus during the packaging, storage, and transportation process in the exporting country, resulting in it being transmitted to people via the environment.

As a new infectious disease, the population has no immunity and is generally susceptible to the new coronavirus. The age distribution of patients across the country showed that people of all ages are not resistant to the new coronavirus and can be infected as long as they fulfilled the transmission conditions. An analysis of 4,021 confirmed patients across the country (diagnosed as of 26 January) also showed that people of all ages were generally susceptible, of which 71.45% are patients were aged between 30 and 65, with 0.35% being children under 10 years old. The elderly and people with underlying diseases such as asthma, diabetes, and heart disease may have an increased risk of contracting the virus. The close contacts of patients, who are infected or asymptomatic, are high-risk groups of new coronavirus infections.

Fortunately, most coronavirus patients present normal or mild symptoms, and their fatality rate is lower than SARS and MERS.

2.4.2 New Coronavirus Vitality

Viruses are different from bacteria. They lack an independent metabolic mechanism, cannot replicate themselves, and can only use the metabolic system of living host cells to assemble and reproduce through nucleic acid replication and protein synthesis. Simply put, viral replication still requires more tools. Therefore, its survival time is limited after leaving the host cell. It will also change according to the environmental conditions.

In 2003, scientists researched the survival and resistance of SARS in the external environment and objects by simulating different environments, objects, and humidity. The results proved that SARS is not resistant to dry conditions as its survival time under dry conditions is relatively short. If it is kept in a liquid state, it remains infectious for a long time.

Although the survival time of SARS differs in different dry states, it has a strong ability to survive in the external environment. The coronavirus, which belongs to the same coronavirus type as SARS, has a sequence of close to 80% homology and is transmissible in vitro.

On 17 March 2020, the National Institute of Allergy and Infectious Diseases (NIAID) published its research results on the in vitro infectivity of the new coronavirus in the *New England Journal of Medicine*. Researchers evaluated its stability by simulating the virus deposition caused by infected persons coughing and touching objects in a home or hospital environment. Tests showed that when a patient with the new coronavirus coughs or sneezes, the virus leaves the body by droplets. When a patient takes a deep breath or talks, he produces aerosols containing the virus.

During the 3-hour experiment, surviving new coronaviruses were detected in aerosols, with their infectivity decreasing over time. The viral concentration per liter of air decreased from 103.5 (TCID50, the lowest concentration to result in 50% cell infection) to 102.7. Its half-life is 66 minutes, that is to say, the active virus would be reduced by half every 66 minutes. During these 3 hours, it remained highly infectious.

The researchers detected that the virus could survive stably for longer periods on the surface of objects. Its infectivity is no longer detectable on copper surfaces after 4 hours, while it disappears after 24 hours for cardboard surfaces. On stainless steel and plastic surfaces, its survival can be up to 48 hours and 72 hours respectively.

Although the virus can survive in vitro, its actual infectivity depends on its viral concentration. On the surface of inanimate objects, its number decreases relatively quickly, while its number in aerosols decreases very slowly in closed environments.

The higher the viral concentration, the higher the risk of infection. This is also the rationale behind indoor ventilation to prevent pandemics. Increasing air circulation can reduce the viral concentration of aerosols, thus reducing the risk of infection.

As the new coronavirus survives longer on smooth and non-porous surfaces, factors that can destroy its structural stability will cause it to inactivate, such as high-temperature treatments above 56°C for 30 minutes and ultraviolet radiation for 1 hour. In contrast, the effects of chemical substances are more direct. For example, washing hands with soapy water, 0.5% hydrogen peroxide, or 62–71% alcohol to disinfect items, can inactivate the virus within a minute.

2.4.3 How to Detect COVID-19 transmissions?

The only way to get an accurate diagnosis of a new coronavirus infection is through a specialized testing method via a detection kit that is further divided into nucleic acid and antibody detections.

The principle of nucleic acid detection is real-time quantitative PCR (polymerase chain reaction), which can quickly amplify a small amount of nucleic acid. After obtaining the nasopharyngeal swabs, sputum, alveolar replacement fluid, and parts that may contain the virus from potential patients, it finds the nucleic acid of the virus within to detect the virus. However, in reality, normal human cells, as well as other bacteria and viruses present in the sample, may interfere with the test results. Therefore, the required nucleic acid fragments are specifically amplified by the PCR strategy. When a large number of samples are amplified, the detection result can be accurately assessed.

PCR can be detected at the initial stages of infection. However, due to the contradictions between high standards and high demand, they naturally lead to insufficient early detection capabilities. The characteristics of RNA viruses make it easier to mutate its characteristic genes, which to some extent, leads to false negatives by the early detection kit.

The target of antibody detection is the immune globulin produced in the human body to fight the virus. When a new virus invades the human body, the immune system is immediately activated. The first to kick in is a class of antibody molecules called immunoglobulin M (IgM), which binds to the protein on the virus surface to deactivate and mark it for destruction by macrophages. A few days later, the system will produce a second antibody-immunoglobulin G (IgG) to continue the fight. IgM has a short lifespan and disappears after staying in the blood for three to four weeks. But the immunity formed by IgG is much longer and may last for many years or even a lifetime.

There are three signs of antibody tests that require attention. A single positive IgM means that the person was infected recently or perhaps, currently. When IgM and IgG are positive, it means that the user was infected at some point in the past month. A single IgG positive means that the infection occurred more than a month ago, so users should now be immune to the infection. A negative result may mean that you are not infected, or that the infection is too early for antibodies to appear. This is because IgM usually does not appear until 7 to 10 days after infection.

The antibody test kit only requires a patient's blood to be extracted and dripped on the test paper, and more accurate results can be obtained in 5–10 minutes. However, in principle, an antibody test requires the human body to produce a certain immune response first before it can be successfully detected about 14–21 days after infection. On the other hand, nucleic acid testing directly determines if the virus is present in the patient, and it can be successfully tested 1–2 days after infection.

Regardless of nucleic acid testing or antibody testing, it is only with large-scale testing of the population that sufficient data can be obtained to make correct decisions.

3

Immune Totem

3.1 How Much Do You Know About the Human Immune System?

The human immune system is the body's guardian and an important system that controls immune responses and functions. Consisting of immune cells, organs, and substances, it recognizes and eliminates antigenic foreign bodies, coordinates with other systems, and maintains the stability of the body's environment and physiological balance. It is also the most effective weapon in preventing pathogens from invading. It can detect and remove foreign bodies and pathogenic microorganisms, as well as factors that cause internal environmental fluctuations, and it can maintain the health of the human body.

3.1.1 Immune System Performing Its Function

The distribution and division of the body's immune cells have an important strategic significance. In terms of cell functions, it is divided into innate and adaptive immune cells.

Innate immune cells

Innate immune cells include mast cells (blood cells), macrophages, neutrophils, natural killer (NK) cells, and dendritic cells, as well as others.

Mast cells are "sentinel" cells guarding the body's portal. They are mainly distributed in the skin, submucosal tissues, and surrounding walls of blood vessels where microorganisms must pass through. They can recognize various danger signals unique to microorganisms, release inflammatory mediators in cytoplasmic granules, gather various immune cells to the invaded tissues and initiate the inflammatory process.

Macrophages and neutrophils are collectively called phagocytes. Macrophages are "resident border troops" distributed in various tissues throughout the body. They have strong phagocytic and killing abilities and are the first line of defense for microorganisms after passing through the body surface.

Neutrophils, which account for 2/3 (60–70%) of the white blood cells in the peripheral blood, are "field" troops that constantly patrol the body as blood circulates and can penetrate blood vessels under the chemotaxis of chemotactic media (biochemical substances) before rapidly reaching the infected tissue and performing the function of phagocytosis of microorganisms. They have a lifespan of only a few days, so they are also called the immune system's "death squad."

Natural killer cells are important immune cells in the body. They are not only related to anti-tumor, anti-viral infections, and immune regulations, but in some cases, they are also involved in the occurrence of hypersensitivity and autoimmune diseases. They are lymphocytes without the typical T and B lymphocyte surface markers and characteristics, primarily generated from hematopoietic stem cells and mature in the bone marrow.

Dendritic cells are usually distributed in a small amount of skin (mucosa) parts in contact with the outside world, primarily the skin (those on the skin are called Langerhans cells) and the inner lining of the nasal cavity, lungs, stomach, and intestine. Immature dendritic cells can also be found in the blood. When they are activated, they move to lymphoid tissues and interact with T and B cells to stimulate and control the appropriate immune response.

Most dendritic cells in the human body are in an immature state and with low expression of costimulatory and adhesion molecules. The ability to stimulate an in vitro proliferation response of mixed lymphocytes is low, but immature dendritic cells have a strong antigen phagocytic ability and they undergo differentiation into mature dendrites when they ingest antigens (including in vitro processing) or are stimulated by certain factors. They express high levels of costimulatory and adhesion factors. In the process of maturation, these cells migrate from peripheral tissues in contact with antigens into secondary lymphoid organs, contact T cells, and stimulate immune responses. Dendritic cells, as the most powerful antigen-presenting cells found so far, can induce the production of specific cytotoxic T lymphocytes (CTL). Studies in recent years have shown that the application of tumor-associated antigens or antigen polypeptides to shock sensitized dendritic cells in vitro, as well as reinfusion or immunization in tumor-bearing hosts, can induce specific cytotoxic T lymphocytes anti-tumor immune responses.

Adaptive immune cells

During the evolution of animals from basic forms to advanced, their immune system has also changed from a simple to a more complex and effective one. An adaptive immune system with high memory and memory functions has appeared in invertebrates. The "adaptability" in this case means that the immune system changes its state after being stimulated by microorganisms or other foreign substances in the living environment and acquires immunity against microorganisms or antigens to effectively completing the mission of defense.

In fact, the adaptive immune system is leap-based on the innate immune system, which adds "modernized" components and functions. One of its main features is the ability to distinguish the subtle differences between different microorganisms or antigens, which seem to be somewhat similar to the "precision strike" of modern warfare. Hence, it is also called the specific immune system.

The T and B lymphocytes are the "modernized" army of the immune system. The total number of lymphocytes in the human body is equivalent to the number of brain cells or liver cells. They use lymph nodes as their "camp" and circulate continuously between the blood and lymphatic system.

The T and B cells recognize antigens through the T cell receptor (TCR) and B cell receptor (BCR) expressed respectively. The adaptive immune response (reaction) can be divided into cellular and humoral immune responses.

Cellular immunity

The cellular immune response is a complex continuous process, which can be roughly divided into three stages, namely induction, reaction, and effect. Specifically, after T cells are stimulated by antigens, they proliferate, differentiate, and transform into sensitized T cells (also known as effector T cells). When the same antigen reenters the body cells, the direct killing effect of sensitized T cells (effector T cells) on the antigen and synergistic killing effect of cytokines is released by sensitized T cells, which is called cellular immunity.

The cellular immune mechanism includes two aspects. Firstly, the direct killing effect of sensitized T cells. When the sensitized T cell and target cell with the corresponding antigen come into contact again, they specifically bind to produce a stimulating effect, change the target cell membrane's permeability, cause a change of osmotic pressure in the target cell, further resulting in its swelling, dissolving and death. In the process of killing target cells, sensitized T cells are unharmed and can attack other target cells again. Sensitized T cells involved in this effect are called killer T cells.

Secondly, this process coordinates with lymphokines to kill target cells. For example, skin response factors can increase the permeability of blood vessels and make it easier for phagocytes to swim out of the blood vessels. Macrophage chemokines can attract corresponding immune cells to the antigen site to facilitate phagocytosis, killing, and eliminating the antigen. Due to the synergistic effect of various lymphokines, the immune effect is expanded, and the purpose of eliminating antigens and foreign bodies is achieved.

Humoral immunity

The humoral immune response, like the cellular immune response, is divided into three stages.

In the induction phase, upon entering the body, most of the antigens, except for a few which can directly act on lymphocytes, must be ingested and processed by phagocytes. The processed antigens can expose the epitopes hidden within. Thereafter, the phagocytes will present antigens to T cells, stimulating them to produce lymphokines, which in turn stimulate B cells to further proliferate and differentiate into plasma cells and memory cells. A small number of antigens can directly stimulate B cells.

During the reaction phase, B cells begin a series of proliferation and differentiation to form effector B cells after undergoing antigen stimulation. In this process, a small portion of B cells become memory cells, which can retain the memory of the antigen months or even decades after it disappears in the body. When the same antigen enters the body again, the memory cells will proliferate and differentiate rapidly to form a large number of effector B cells and produce a stronger specific immune response to clear the antigen in time thereafter.

During the effect phase, the antigen becomes the target of action. The antibody produced by effector B cells can specifically bind to the corresponding antigen to induce an immune effect. For example, the binding of antibodies to invading germs can inhibit the reproduction of germs or adhesion to host cells, thereby preventing the occurrence of infections and diseases. When antibodies are combined with viruses, they can enable the virus to lose its ability to infect and destroy host cells. In most cases, further changes will occur after the binding of antigens and antibodies, such as the formation of precipitates or cell clusters, and they undergo phagocytosis thereafter.

3.1.2 Natural Immunity and Acquired Immunity

The immune system is divided into two categories, namely innate immunity and acquired immunity. Innate immunity is acquired from birth, while acquired immunity is gradually acquired during the survival process.

Innate immunity, also called non-specific immunity, is the body's innate function of maintaining health. It instinctively rejects and carries out phagocytosis of foreign substances, germs, foreign bodies, and other factors. It includes body, blood-brain, and blood-fetal barriers, cell phagocytosis, antibacterial substances in body fluids and tissues of the human body tissues.

The non-specific immune function consists of three lines of defense.

The first line of defense: Mechanical barriers, such as skin and mucous membranes, are responsible for preventing pathogenic microorganisms from entering the body.

The second line of defense: Phagocytic cells, which exist in the blood and various tissues, are used to swallow and destroy pathogenic microorganisms such as bacteria and viruses entering the body.

The third line of defense: A variety of antimicrobial substances is present in the blood, tissue fluid, and various secretions. For example, the lysozyme in saliva can dissolve bacteria entering the oral cavity. After human cells are infected by a virus, they can produce interferon to kill the virus. It can be seen that non-specific immunity is the basis of specific immunity, and specific immunity and non-specific immunity complement each other and jointly maintain our health.

Take bacteria as an example. If a certain type of bacteria enters the human body from the respiratory or digestive tract, the respective tracts will block it first. If it is not blocked, it enters the blood or tissues, and the phagocytes fight against it — by eating or destroying it. If it has not been eliminated, it will enter the lymph nodes and spleen, where T cells become sensitized to T cells upon stimulation, and B cells produce antibodies upon stimulation and continue interacting with it. If the bacteria invade again in the future, more sensitized T cells with recognition function will be produced. B cells will also produce more antibodies, making the body more powerful to fight against it.

Acquired immunity, also known as specific immunity, is a certain specific immunity acquired after taking vaccines, vaccinations, or from suffering from certain diseases after exposure to pathogenic microorganisms, which means that immunity is formed against a certain disease. For example, there is immunity to hepatitis after suffering from hepatitis as anti-hepatitis antibodies are produced in the body under the stimulation of hepatitis pathogens.

The building of specific immunity begins from the bone marrow.

Bone marrow is an immune organ. There is a very important cell in the bone marrow called hematopoietic stem cells, which have varying differentiation potentials. According to the body's needs, stem cells can differentiate into red blood cells, white blood cells, and phagocytes, some of which will become lymphoid stem cells.

Lymphatic stem cells are divided in two ways. The first one will be for them to stay in place and turn into B cells in the bone marrow, before reaching and settling in with the spleen and lymph nodes with the blood. The other way will be for them to reach the thymus with the blood, differentiate into T cells within, and arrive in the spleen and lymph nodes in the blood, while awaiting alongside B cells to prepare to perform their immunity functions.

After external pathogenic microorganisms (antigens) invade the body, if they can overcome the various barriers, enter the blood, and reach the spleen and lymph nodes, T and B cells should be able to play their roles.

First, T cells are stimulated by this pathogenic microorganism to enter an activated state and become sensitized T cells with two characteristics, namely the ability to attack this pathogenic microorganism and kill it, as well as to recognize and remember it so that they can attack it the next time they see it.

B cells will then produce a substance when stimulated by pathogenic microorganisms. This substance, an antibody, can bind to pathogenic microorganisms and make them inactive. If this pathogenic microorganism invades the body again in the future, sensitized T cells (which activate T cells) will rush forward to fight against them, while antibodies will bind to them and make them lose their ability to cause diseases.

As such, the body's immune system only has to see a certain pathogenic microorganism once it develops an immunity to it. This is why humans can be immune to a pathogen, as long as they have taken or injected a corresponding vaccine, or have had this infectious disease.

3.2 Immunity Changes Spanning 157 Years

After the COVID-19 outbreak, temperature taking has become a daily routine for many. A swipe to the forehead will display a person's body temperature. However, if you observe closely, you will find that the measured body temperature is mostly in the lower end of 36°C, not 37°C. The common knowledge is that the standard body temperature is around 37°C, which deviates from previous standards. What does this mean to us?

After further research, the rationale behind the decrease in the basal body temperature has finally been revealed. This is a social change and even more an immunity change.

3.2.1 The 37°C We Can No Longer Return to

Human body temperature refers to the temperature inside the human body, for instance, the abdomen, chest, rectum, mouth, and brain, and not the skin. It has always been a key vital sign used clinically to assess health. The human body constantly metabolizes and produces heat, as well as dissipates heat to the environment. They coordinate to maintain a constant body temperature. The constant temperature of animals, including humans, can be regulated in the body's thermoregulatory center in the hypothalamus.

Therefore, the normal body temperature is not determined by cell metabolism and muscle movement, but rather, it has been naturally "designed."

For more than a century, 37°C has been regarded as the standard human body temperature. An obvious mark can be seen at the 37°C mark on commonly used household thermometers intended to remind us of it.

In fact, people previously did not know what the average or normal body temperature of than a hundred years ago. In 1851, a German physician named Karl Ondrich obtained millions of axillary temperatures from 25,000 patients in Leipzig, thus establishing 37°C as the standard of a normal body temperature.

With medical technology developments and continuous research, we started having a more in-depth understanding of body temperature. People gradually learned that human body temperature is not constant and fluctuates throughout the day. Studies showed that it is the lowest from 2 a.m. to 5 a.m. and the highest at 5 p.m. to 7 p.m., with the variation range at 0.5°C – 1°C. For long-term night workers, their body temperatures rise at night and drop during the day. In addition, body temperature varies according to one's gender. The average body temperature of women is 0.3°C higher than men. The body temperature of women rises slightly during the premenstrual and early pregnancy periods, while it is lower during the ovulation period. Age is also one of the factors affecting body temperature. The body temperature of newborns is susceptible to external changes, and their body temperature can be slightly higher than adults.

However, "the normal human body temperature is 37°C," that was once regarded as the scientific definition, has been continuously questioned in recent years. In 2017, a British study took 250,000 body temperature records of 35,000 subjects and found that the average body temperature was around 36.6°C.

On 7 January 2020, Stanford University School of Medicine professor Julie Parsonnet and her team published a study in the scientific journal ELIFE. They found that since the 19th century, the average adult body temperature has continued falling from 37°C to 36.6°C, a decrease of 0.4°C in less than 200 years.

Many also believed believe that the difference in average body temperature was due to increasingly precise measurement equipment and more scientific measurement methods. Hence, as compared to the 19th century, has our "standard basal body temperature" really changed?

In this study, Julie Parsonnet's team analyzed human body temperature data sets from three different periods, including the medical records of American Civil War veterans (1862–1930), the National Health and Nutrition Examination Survey (1971–1975), and Stanford University Translational Research Comprehensive Database Environmental Research Group (2007–2017).

This data comprised nearly 680,000 body temperature measurements, spanning 157 years and 197 birth years.

The comparative study showed that the average body temperature of men in the 19th century was 0.59°C higher than in modern times and decreases by 0.03°C every decade. Since 1890, the body temperature of women has also dropped by 0.32°C and by about 0.029°C every 10 years. Professor Julie Parsonnet also said that this is unlikely to be a systematic error in the measurement data, but a real physiological difference.

3.2.2 Immune Changes due to Body Temperature Changes

The temperature of 37°C is finally a thing of the past. So, what caused the decrease in body temperature? Current research is unable to provide a definite answer. However, these researchers speculated and analyzed the possible factors.

Professor Julie Parsonnet believed that this may be related to inflammation in the body, which can produce various proteins and cytokines, as well as increase the body's metabolism rate and temperature.

For example, in the mid-19th century, 2–3% of the population may have had active tuberculosis. This data is consistent with the data of Julie Parsonnet's team investigating active tuberculosis among American Civil War veterans from 1862 to 1930. The survey showed that there were 737 cases of active tuberculosis (3.1%) among 23757 subjects and confirmed that those with tuberculosis or pneumonia had a higher body temperature of 0.19°C and 0.03°C respectively.

In 2008, healthy volunteers from Pakistan participated in a mini-study. In Pakistan, there is a high risk of chronic infections, but its temperature is closer to the value reported by Dr. Ondrich more than a century ago (the average, median, and mode were 36.89°C, 36.94°C, and 37°C respectively), which also proved that inflammation results in an increase in body temperature.

During Dr. Ondrich's era, the life expectancy of humans was only about 40 years old, and most were affected by untreated chronic infections such as tuberculosis, syphilis, and periodontitis. In the past 200 years, due to advances in medical standards and an improvement in sanitation, food availability, and living standards, tuberculosis, malaria, and chronic infections caused by trauma were reduced, alongside an improvement in dental hygiene. The advent of antibiotics also helped to reduce chronic inflammation. A study of participants in the National Health and Nutrition Examination Survey of the United States from 1971 to 1975 showed that between 1999 and 2010, the abnormal C-reactive protein levels decreased by 5%. The C-reactive protein (acute proteins) is known to increase dramatically in response to infection and tissue injury.

In addition, it may be related to changes in the living environment, including indoor temperature, microorganisms in the body, and food. Since people intentionally and unintentionally ingest genetically modified foods at varying levels, the modified genes entering their bodies conflict with the immune cells to some extent, albeit most being unaware in most cases. However, some experience obvious reactions, with allergies being the most common. In general, due to an environmental change, improvement in life comfort levels, intake of genetically modified food, lack of exercise, the introduction of drugs, and other factors, it has become the main reason for changes in human body temperature we see today.

For example, if the environmental and basal body temperatures vary greatly, the human body will consume more energy to maintain a normal body temperature, which stimulates the body's metabolic vitality to a large extent. The most significant is the resting metabolic rate of the human body, which refers to the total calories consumed by the human body to maintain its body functions during rest. It is also the largest component of typical energy expenditure, accounting for approximately 65% of the daily energy expended by sedentary individuals. However, maintaining a constant body temperature with a fluctuating ambient temperature can consume up to 50–70% of the daily energy intake. When the ambient temperature is lower or higher than what humans can maintain at a normal temperature with the lowest energy consumption, the body's resting metabolic rate will increase.

Another example would be the unstable heating systems and lack of air-conditioners in the 19th century. By the 1920s, the heating system covered most of the population. Today, comfortable indoor temperatures are commonplace. This also means that with technological developments, the increase in comfort levels of humans will weaken the body's metabolic vitality to a certain extent as it enables us to maintain a normal body temperature with minimal energy consumption, alongside lesser changes in body temperature, thus leading to a decrease in body temperature.

Humans have undergone physiological changes in the past 200 years, and the average body temperature is 1.6% lower than before industrialization. Will this change affect humans?

It is well understood that body temperature is linked to basal metabolic rate. For every 1°C increase in body temperature, basal metabolism will increase by 13%, while human immunity will increase by 500–600%, which is 5–6 times. Research by Arturo Casa de Farr, MD of Johns Hopkins University, also confirmed that between 27°C and 40°C, for every 1°C increase in temperature, 6% of the fungus loses its ability to infect the host. This is why there are tens of thousands of fungi species that can infect reptiles and amphibians and other temperature-changing animals, but only a few hundred species can threaten humans and other mammals. The bat, known as the King of Poisons, remains safe even though it can carry hundreds of viruses, which may be fatal to humans. The primary reason is due to the powerful immune system created by its body temperature of 40°C.

When a person's body temperature dips below 37 degrees, it means that the body's immune system's ability to fight against bacteria and viruses is reduced. Generally, we encounter a large number of bacteria and viruses every day. Even if there are fungi, bacteria, or viruses invading the human body unintentionally, white blood cells can find abnormalities and destroy pathogens in the body at a quicker rate when the body temperature is high and blood flow is fast. On the contrary, white blood cells become less efficient when there is a low basal metabolic rate, alongside a decrease in body temperature and blood flow rate. This also means an increased probability of disease.

According to research estimates, there are more than 380 trillion viruses in a human body, and these groups are collectively referred to as the "human virus group." However, they are not the dangerous viruses that you often hear about. Many can infect bacteria living in humans. They are called bacteriophages, which can kill bacteria. They will take over the bacterium's cellular machinery and force them to produce more phages instead of bacteria.

Most viruses are resistant to cold temperatures, not heat. While they can survive well at temperatures below 0°C, they are inactivated within a few minutes to ten minutes at 55°C–60°C. A virus, composed of genetic material and some proteins, does not have a complete cell. Since the outbreak of the new coronavirus, the talk about alcohol disinfection by experts shows that it can dissolve the protein membrane on the virus surface and deactivating the gene chain within, hence there is no way for it to infect humans. Ultraviolet rays destroy viruses by using the energy of ultraviolet rays to penetrate the protein membrane of the virus and directly act on its genetic material.

Body temperature change has been ongoing for more than 100 years. Even if it continues dropping, it has an adaptive mechanism, a physiological mechanism for humans to adapt to nature. However, what is called accidental hypothermia in Western medicine is a typical manifestation of Qi deficiency in Traditional Chinese Medicine.

With rapid developments in science and technology in an era of changes, humans have led comfortable lives but also unknowingly paid a heavy price. Human immunity has not improved with the increased level of comfort. On the contrary, it has shown a downward trend when there is an increased level of comfort and drug abuse. This slight dip of 0.5–1°C has undermined the human immune system, resulting in a reduced ability of humans to fight bacteria and viruses.

In the first chapter of *Suwen (Questions Of Fundamental Nature)* titled *Shanggu Tianzhen (The Universal Truth) of Huangdi Neijing (Inner Canon of the Yellow Emperor Bo)* replied that ancient people, who knew the importance of strengthening the immune system, were able to follow the principles of yin and yang, adapt accordingly and reach the standards. They also had a moderate diet, regular work, and rest, and avoided excessive sexual intercourse so that they could maintain their vigor and live to a ripe old age.

Immunity changes as a result of a change in body temperature prompt us to rethink societal changes. After the pandemic, people should think more about their current lifestyles and make adjustments, such as getting more sun exposure, soaking their feet, enduring moderate hunger, adjusting food structure (food is divided into negative yin and positive yang in Traditional Chinese medicine; a more positive food intake can help increase yang and body temperature) or doing more aerobic exercises to repair and improve one's yang, as well as to regain the disappearing 0.5–1°C body temperature.

3.3 Aging of the Immune System

Aging is a complex physiological process as it involves time and human-environmental interactions, resulting in changes in the structure and functions of molecules, cells, and the body. It is characterized by the progressive decline in physiological functions and ability to stabilize the internal environment of tissues, which will lead to an increase in the incidence of degenerative diseases and death.

In the 1950s, Walford conducted pioneering research in biogerontology and pointed out in Immunological Theory of Aging that a decline in immune functions may cause aging in 1969. The concept of "immunosenescence," first proposed by him, was further explored by academia. The aging of the immune system is not only the inevitable result but also an important cause of body aging.

3.3.1 Natural Immunity During Aging

During aging immunity, the body's innate and adaptive immune systems will be affected. Existing research shows that the impact of aging is greater on the adaptive immune system as compared to the innate immune system, which is reflected by changes in expression of cellular senescence-related markers, secretion of cytokines, cell subsets, and cell function defects. At the same time, the impaired immune surveillance function will accelerate the accumulation of senescent cells and hasten the aging process.

From the perspective of changes to the innate immune system functions due to aging, a significant number of studies showed that these changes begin with a decrease in the barrier function of the epithelial layer, gastrointestinal tract, and respiratory mucosa, alongside a decline in immunoglobulin levels.

With the innate immune system being the body's first line of defense, the main characteristic of skin aging is a damaged protective function "barrier" structure due to the decreasing hair quantity and coverage. The skin's elasticity decreases with the lower number of sebaceous glands, which severely impairs the skin's immune defense capabilities.

In the innate immune system, neutrophils can produce hydrogen peroxide, chloride, and peroxidase to form a myeloperoxidase (MPO) bactericidal system. At the same time, it has powerful phagocytosis and killing effect on pathogens with the synergistic effect of complement fragments of antibodies.

The expression of CD16Fcγ receptors has decreased even though the number of neutrophils has not decreased among the elderly. As a result, the phagocytosis of superoxide produced by Fc receptors is affected, indicating that the immune function of neutrophils in the elderly is adversely impacted by the decrease in Fc receptor binding.

In addition, for natural killer (NK) cells which play a monitoring role, although there was an increase in NK cells among the elderly, their toxicity decreased, while their antibody-dependent cellular cytotoxicity remains constant when measured by the level of cytokines and chemokines produced by each cell.

The lethality of senescent natural killer cells is reduced, and the expression of T-bet and Eomes is also significantly reduced. Under the stimulation of IL-2, senescent natural killer cells secrete insufficient IFN-γ and IFN-α but they also secrete more IL-1, IL-4, IL-6, IL-8, IL-10, and TNF-α. Studies have shown that changes in the toxicity of natural killer cells are related to the imbalance of zinc in the elderly, and the function of natural killer cells can be significantly improved after zinc supplementation.

3.3.2 Adaptive Immunity in Aging

The lymphoid T and B cells, which possess specific recognition functions, play key roles in the three phases of the adaptive immune response. Among them, T cells are particularly vulnerable to the aging process.

The T cells are produced in the thymus and can be divided according to the receptors which bind to them into two forms, namely the CD4+ and CD8+. The ratio of these two cell subtypes shows a certain trend during the aging process. The number of CD8+ cells is increasing during the aging process, while the CD45RA and CD45RO produced by CD4+ and CD8+ T cell expressions are mutually exclusive. The first phenotype can recognize primitive T cells, while the second phenotype can activate T cells.

Studies have shown that the reduction of primordial lymphocytes may be the result of thymus degeneration and chronic antigen stimulation, which also reveal the reason behind the elderly's reduced ability to resist new infections. In addition, during the aging process, primitive T cells show a variety of characteristics such as shortened telomeres, reduced IL-2 production, and the reduced ability to differentiate into effector cells.

The expansion of effector "memory" cells and vaccination results in the clonal expansion of T cells CD8+, CD45RO+, and CD25+ for about 30% of the elderly. The loss of the number and function of primitive T cells is compensated, alongside the production of IL-2 and protective humoral immunity.

Further changes in T cells involve a damaging response to oxidative stress, leading to increased susceptibility to induce cell death and calcium flow dynamics. Current research points out that the reduction of miR181 (microRNA precursor) during aging reduces the T cells' ability to recognize antigens.

Regulatory cells, also called Tregs, are transcription factors with recognition function, which produces subtypes expressing high levels of CD25 and FOXP3. The number of CD4+, FOXP3+, and lymphocytes in elderly patients increases. The accumulation of these cells plays an important role in activating the defense mechanism against chronic infection. At the same time, changing the T17/Treg ratio will result in the body's inflammatory response to inflammation or autoimmune diseases.

B lymphocytes are produced in the bone marrow and mature in the spleen. B cells are responsible for the secretion of antibodies and play an important role in humoral immunity. During the aging process, the number of B cells produced by the bone marrow, and the diversity of receptors decreases significantly, alongside a reduction of antibodies produced. This reduces the body's responsiveness to infection and vaccination and increases the production of autoreactive antibodies.

Studies have shown that given the same antigen intensity stimulation, the number of B cells mobilized is only 1/10 to 1/50 of normal adult animals. For example, the seropositivity rate for the 60 or 74-year-old group after influenza vaccination is 41%–58%, while the positive and protection rates for those over 75 years old decreased to 29%–46%. The age-related changes in the cell composition of the B-cell series are the main reasons for poor antibody response during vaccination and infection in the elderly.

In addition, the increase in memory B cells in the elderly may be related to an increase in inflammatory aging and chronic inflammatory diseases among them. At the same time, aging is accompanied by the tendency to produce Th2 cells in the immune response, alongside the secretion of excessive Th2 cytokines, which may enhance B cell-mediated autoimmune diseases.

3.3.3 Is the Weakening Immunity During Aging Reversible?

Although aging can cause severe immune dysfunction and reduce the vaccine's reactivity, researchers believed that this was an irreversible defect in the immune system for a long time. In 2020, a research team led by the Cincinnati Children's Hospital of the University of Cincinnati School of Medicine in the United States overturned this conclusion in a research paper published in the journal *Science*.

Studies have found that the immune system of the elderly is not weakened, but rather suppressed by the activity of a group of immune cells. Researchers call it the follicular helper T cells, which produce interleukin 10 (Tfh 10), and such inhibition is reversible. The data showed that the release of this suppression, driven by a set of key cells in the human immune system, can trigger a strong response to vaccination among the elderly. Researchers call them "Tfh10" cells, which represent the follicular helper T cells producing interleukin 10 (IL-10).

As mentioned above, aging was previously thought to cause irreversible immune dysfunction. At the same time, aging has persistent low-level immune activation (so-called "inflammation") characteristics and is accompanied by high levels of IL-6. However, the new study found that apart from the increase in IL-6, the serum IL-10 (an effective anti-inflammatory mediator) in elderly individuals also increased, while limiting the body's protective response to pathogens.

In the study, the researchers conducted dozens of different experiments, in a bid to investigate the reasons behind the poor levels of IL-10, and eventually traced the excess IL-10 production back to the cell type they called Tfh 10. Subsequent mouse models confirmed that simple blocking of IL-10 during vaccination can restore the antibody response to nearly the level of young animals.

Of course, this research reflects the combination of advanced computing and laboratory work, many of which were confirmed in experiments involving mouse models and human cells, but more research is still required to prove that Tfh 10 cells can be securely managed in human bodies.

3.4 Sleep and Immunity Under the Motivation of Health

The importance of sleep to the human body is known by all. Among them, the most important and frequently mentioned point is its effect on immunity. Generally speaking, a good sleep state will enhance immunity and resist the invasion of various diseases and pathogens. Long-term insomnia will cause a decline in immunity, and the human body will be susceptible to viruses and bacteria, which will cause various diseases.

Sleep is a rather complex process of physical and psychological changes. In comparison with wakefulness, many physiological functions, including immunity, change during sleep. This cyclical change results in sleep and immunity affecting one another for a prolonged period and determines the body's daily routine.

3.4.1 Sleep is not Easy

A major part of a person's life is spent on sleep, a naturally recurring state of mind and body. It is characterized by changes in consciousness, a decrease in responsiveness to stimuli, and inhibited voluntary movement. A single test cannot detect sleep, and it generally requires multiple physiological and behavioral indicators to measure, including objective changes in electroencephalogram (EEG), electromyography (EMG), or breathing rate.

Sleep is divided into rapid eye movement (REM) sleep, characterized by the rapid movement of the eyes and dreams, and non-REM, the opposite of REM sleep, which lasts for a long time. Studies have shown that REM sleep plays a repair function on the brain, including a series of immune-related tissue cells existing in the brain.

In addition, non-fast eye movement sleep includes deep sleep (SWS), also known as slow-wave sleep. It is characterized by the presence of synchronized EEG and slow waves, as well as incremental wave activity at times. Deep sleep is often regarded as a time to rest, and most people do not dream during deep sleep.

Insomnia is manifested as difficulty in sleeping or maintaining sleep, waking up early, interrupted or non-restorative sleep, and other daytime-related dysfunctions. Symptoms persisting at least three nights a week and lasting for three months or longer are called insomnia. Publicly available information has revealed that the average sleep time of Chinese people is usually about 6.92 hours, with the proportion of deep sleep being less than 1/3. The proportion of Chinese people with insomnia is as high as 36.1%, which is higher than the global figure of 27%.

University of Pennsylvania researchers have demonstrated the close relationship between sleep and the immune system through two experiments involving fruit flies. The first involves the use of Serratia marcescens and Pseudomonas aeruginosa to infect fruit flies. The experimental group is sleep deprived upon confirming that they are successfully infected. The results showed that this group, as well as the control group without sleep deprivation, had acute sleep reactions of varying degrees. As the experimental group had a longer sleep time after infection, their survival rate was higher than the control group.

The second used genetic means to manage sleep. The researchers used the drug RU486 to induce the expression pattern of ion channels and adjust sleep patterns by altering the activity of mushroom body output neurons in Drosophila. The results showed that the survival rate of fruit flies with prolonged sleep time due to induction was higher. Their bodies were also able to clear bacteria faster and more effectively. This proved that increased sleep can improve the immune capacity, anti-infection ability, as well as recovery and survival rates after infection for fruit flies.

3.4.2 How does Sleep Regulate Immunity?

As early as 50 years ago, researchers began to believe that sleep is directly related to the immune system, but there were no effective means to prove this at the time. The development of modern medicine has clarified the close relationship between sleep and the immune system.

The first will be the impact on NK cells, an important immune cell of the body. It is not only related to anti-tumor, anti-viral infection, and immune regulation, but also participates in the occurrence of hypersensitivity and autoimmune diseases in some cases, and can recognize target cells and weaken the medium.

A large number of studies have proved that sleep deprivation reduces the phagocytic ability of white blood cells and lethality of NK cells, and the reduction is proportional to the duration of sleep deprivation. It can gradually recover after returning to sleep. Some researchers have found that long-term insomnia reduces the activity of natural killer cells by depriving the human body of sleep. This convincingly demonstrates that there is a certain correlation between sleep deprivation and NK cell activity.

It was also found in clinical trials that NK cells were reduced by 31% and 37% after 24h and 48h of sleep deprivation in healthy men, and returned to the original active level after returning to normal sleep hours. It can be seen that the change of NK cell activity is positively correlated with sleep deprivation, and it can gradually return to normal after returning to normal sleep hours, suggesting that the immune system is related to sleep.

Secondly, studies on the effect of sleep on lymphocytes showed that while sleep deprivation has no significant effect on the number of T lymphocyte subgroups, it reduces the function of lymphocyte subgroups. This decrease is directly proportional to the sleep deprivation time, and their functions can gradually recover after resuming sleep, indicating that this immune change is related to sleep.

Through sleep deprivation in humans, researchers found that there was decreased proliferation response of peripheral blood lymphocytes to mitogen stimulation, as well as a decreased phagocytic function of multi-row nuclear granulocytes, which is a reduction of the body's cellular immune functions. Sleep deprivation can inhibit the proliferation of splenocytes induced by concanavalin (ConA) and lipopolysaccharide (LPS). The proliferation of splenocytes induced by ConA reflects the function of T cells, while the proliferation of splenocytes induced by LPS reflects the function of B cells. This also means that sleep deprivation can also lead to decreased T and B cell functions, as well as decreased cellular and humoral immune functions.

Thirdly, sleep has an important effect on cytokines. Sleep deprivation affects human brain function and causes neurological dysfunction, resulting in the body's stress response, which will activate the thalamus-pituitary-adrenal axis and increase the secretion of glucocorticoid hormones. This hormone exerts a strong inhibitory effect on human immune function and reduces human immune function. Moreover, this hormone will cause the hypothalamus to release corticotropin-releasing hormones, activate natural killer cells, and enhance the production of IL-1, IL-2, and IL-6 cytokines. IL-6 and TNF-α are signs of systemic inflammation, which can lead to diabetes and cardiovascular disease.

Studies have shown that 13 healthy young men, who slept only four hours a night for five consecutive days, activated lymphocytes and increased pro-inflammatory cytokines IL-1β, IL-6, and IL-17. These pro-inflammatory cytokines, which were still active during 2 days of restorative sleep, were accompanied by increased heart rate and increased serum C-reactive protein (CRP). CRP is synthesized in the liver and controlled by pro-inflammatory cytokine IL-6, interference INF-α (INF-α), and IL-1. Therefore, long-term sleep deprivation not only increases IL-17 and CRP products but is also accompanied by increased heart rate symptoms, which are risk factors for cardiovascular disease.

Finally, sleep affects immunoglobulin and its complements. Studies have shown that after 48 hours of sleep deprivation, immunoglobulin IgG and IgM increase, while IgA decreases. After 56 hours of sleep deprivation, serum immunoglobulins IgA and IgM remain unchanged, while IgG and complement C3 decreased, and serum complement CH50 also decreased significantly. GH deficiency can cause thymus and spleen atrophy and decrease the number of lymphocytes.

3.4.3 How does Immunity Regulate Sleep?

Apart from the impact of sleep on immunity, immunity also regulates sleep. It is the long-term interaction between sleep and immunity which maintains the normal operation of the body. Immunity changes the activity of the central nervous system and regulates the process of sleep through the interaction of pro-inflammatory signals generated in the periphery and the central nervous system (CNS).

Firstly, it is the role of cytokines in sleep regulation, such as IL-1 and pathogen-related molecular patterns (PAMPs), as well as lipopolysaccharide acting on the vagus nerve, to project to multiple brain regions involved in sleep regulation, including the nucleus tractus solitarius, ventrolateral medulla nucleus, paraventricular nucleus of the hypothalamus (PVN), and the supraoptic nuclei and amygdala. According to reports, vagus nerve amputation can block the induction of sleep by systemic cytokines, as well as the expression of cytokine mRNAs in the brain induced by system cytokines.

Secondly, many parts of the CNS have macrophages expressing toll-like receptors (TLRs). When these macrophages are activated by PAMPs, they produce inflammatory cytokines such as IL-1, which can diffuse into the brain. In addition, endothelial cells express IL-1 receptors activated by circulating IL-1, leading to the production of local prostaglandin e2 and triggering immune activation in the brain.

Thirdly, the blood-brain barrier (BBB) actively transmits a variety of immunomodulatory molecules in the CNS. This process is actively coordinated by CNS, which is different from uncontrolled leakage by passive diffusion or dysfunction of the BBB. Due to infection and aging, the BBB function also changes, which may be beneficial for the active transport of IL-1, IL-6, and TNF. In these cases, changes in sleep patterns further increase the transport of inflammatory mediators in the BBB.

Finally, when immune cells (usually monocytes) are activated, they can connect to the vasculature of the brain and other vasculature. Peripheral inflammation signals stimulate microglia to produce CCL2 (MCP1), which further attract monocytes to the brain. Astrocytes stimulated by cytokines can also produce CCL2 to attract immune cells to the brain.

3.5 Disorders of the Immune System

Generally speaking, a healthy immune system has three basic standards, which show that killer T cells and antibody-making B cells are balanced, helper T cells and regulatory T cells are balanced to activate or deactivate the immune system, and the immune system can distinguish foreign invaders (such as viruses or bacteria) and parts of the body (such as beneficial bacterial cells and tissues).

When these standards are not in order, autoimmunity problems will arise. The body begins producing too many killer T cells or antibodies (depending on the autoimmune disease) and cannot stop from doing so, resulting in a dysfunctional immune response.

3.5.1 Autoimmune Diseases

Autoimmune disease, also known as an autoimmune disorder, is a disease in which a body's immune system attacks its normal cells. It is an issue in which the normal immunity weakens, but the abnormal immunity accentuates it.

The so-called abnormal immunity recognizes friends as enemies and treats those that are not viruses or bacteria in their body as viruses or bacteria to attack them, hoping to drive them out. Antibodies of the immune system in the human body are originally designed to attack and eliminate foreign antigens or abnormal cells in the body (such as tumor cells), which is a physiological mechanism to protect the body. However, in some cases, the immune system may produce antibodies against normal cells in the body (or even various normal components in the cells), causing abnormal excessive inflammation or tissue damage, thereby affecting health and causing disease.

For people with autoimmune diseases, the immune cells need to attack foreign invaders instead of their own tissues. The name of "autoimmune" disorders covers at least one hundred diseases and not a specific single disease, so it is easily understood.

Unlike various cancer names, autoimmune diseases often carry the word "cancer" and are located where malignant tumors are found. For example, breast cancer is a tumor of the breast, colon cancer is a tumor of the colon, and skin cancer is a tumor of the skin. Since the names of most diseases do not contain "autoimmunity," they may sound like another disease altogether, for instance, Hashimoto's thyroiditis, rheumatoid arthritis, systemic lupus erythematosus, Shogan's Syndrome, multiple sclerosis, and others.

In addition, the names of autoimmune diseases do not indicate the body part where the disease occurred, which is also confusing. Some autoimmune diseases are systemic, that is, they attack tissues throughout the body, for example, lupus erythematosus. Other autoimmune diseases tend to occur in specific organs and only attack specific parts or organs. For example, Hashimoto's disease only involves the thyroid. But whether it is the former or the latter, their names are not a clear indicator of where the problem occurred. Hashimoto's disease and Graves' disease occur in the thyroid, multiple sclerosis occurs in the brain and spinal cord, leukoplakia occurs in the skin, and pernicious anemia occurs in the blood cells. Although the locations of these diseases are different, people now know that the problems behind them are very similar.

In fact, research in recent years has shifted its focus from specific organs affected by these diseases to underlying mechanisms causing these diseases. "The causes of autoimmune diseases are similar" is key to the treatment and reversal of these diseases in academia. More than one hundred different autoimmune diseases have similar characteristics. They are serious chronic diseases and have potential issues for the immune system.

Another common point is inflammation, which refers to inflammation and swelling in the body, including the brain. It can cause a wide range of symptoms, including fatigue, swelling, muscle or joint pain, and abdominal discomforts such as diarrhea, concentration difficulties, or brain fog (a state in which it is difficult for the brain to think clearly).

3.5.2 Inducing Factors of Immune Disorders

By using functional medicine methods and focusing on the main causes of immune dysfunction, researchers have discovered many potential predisposing factors (leading to unhealthy immune responses) of these diseases. Studies have found that many autoimmune diseases are induced by similar factors, such as gluten, stress, beneficial bacteria, and poison.

Gluten

Modern agricultural technology includes genetic modification, which means that cereal seeds such as corn, soybeans, and wheat can be modified in the laboratory to be larger or more efficient in resisting disease. As a result of changing genes of these grains, they produce a protein that did not originally exist. Animal studies have found that these proteins are very difficult to digest and can easily cause heartburn, gastroesophageal reflux, and flatulence after eating.

In addition, there is evidence that these proteins trigger intestinal immune responses that cause autoimmune problems. Autoimmunity means that the cells of the immune system are damaged and attack their own cells instead. Among them is gluten, the protein contained in wheat, barley, Kamut, and spelt. As a result of genetic modification, it makes gluten in grains stronger and more concentrated.

The higher concentration of gluten in food is related to the increase in the proportion of food allergies in the past few decades. This is because gluten is a relatively new ingredient in our diets. During the hunter-gathering times, our ancestors ate animals, nuts, seeds, and berries instead of grains. Then it turned into an agrarian society (only about ten generations ago), with food changing and consumed according to seasons. The advantage of this is that you can keep changing your diet, but eating the same food constantly can easily increase the risk of allergic reactions.

In addition, the problem with gluten is that it is difficult to digest. When a large number of large particles of gluten enter the blood, they alert the immune system, which treats it as a foreign substance, and produces antibodies to attack them. Unfortunately, when antibodies attack gluten, they also mistakenly attack body tissues. This phenomenon is called molecular mimicry, and it is generally believed that gluten causes autoimmune diseases. Molecular mimicry not only targets gluten but also occurs when the immune system, such as self-organized tissue, mistakes it for a foreign object.

Chronic Stress and Hormonal Imbalance

When some people feel too emotionally stressed, they do not eat regularly, can lack sleep, or exercise excessively. These actions will exert pressure on the body and facilitate the adrenal glands' secretion of the stress hormone cortisol. In addition, some only take care of their bodies, but they are burdened with anxieties, worries, depression, or severe and persistent emotional trauma, which can trigger the same cortisol response. The adrenal glands are small glands located above the kidneys that secrete all stress hormones. Of course, not all stress hormones are bad. For example, in an emergency, the adrenal glands secrete cortisol and adrenaline to provide you with the energy to move quickly and seek assistance. Before important speeches, stress hormones can generate energy, as well as help one to focus and think.

However, chronic stress means that the cortisol index continues rising, resulting in a weakened immune system, which makes it difficult to recover. Chronic stress can also cause adrenal fatigue, making the adrenal glands so tired that they cannot produce hormones, including adrenaline, dehydroepiandrosterone (DHEA), and testosterone, which allow the body to function normally. Adrenal fatigue can lead to unexplained fatigue, an inability to wake up after a good night's sleep, a heavy burden in one's heart, a desire to eat sweet or salty food, low blood pressure, low blood sugar, and irritability, among other side effects.

Adrenal fatigue, also called adrenal failure and overworked adrenal glands, is related to inflammation and autoimmune diseases. Therefore, it is very important to understand and appropriately manage life's stressors. Stress harms the number of beneficial bacteria in the digestive tract and may also cause autoimmune diseases. If you always feel tired, get sick frequently, have arthritis, menstrual irregularities, menopausal disorders, and difficulty losing weight, there may be issues with your stress hormones.

An Imbalance of Beneficial Bacteria in the Gut

The body's immune cells, especially killer T and B cells, are key to autoimmune problems. If these cells cannot function properly, the body will start to attack its tissues. To help these cells function better, it is important to understand their developmental process.

After reaching adulthood, the bone marrow produces immune cells, which are then transferred to the thymus (a small organ under the breastbone), lymph nodes, and intestinal-related lymphoid tissues below the surface of the intestinal mucosa. When the fetus is in the mother's womb, the thymus is very active and remains the main area for immune cells after birth. With age, although the thymus still assists in the maturation and development of immune cells, its activity will decrease. The intestinal mucosa must have sufficient beneficial bacteria (gut flora), which plays an important role in helping immune cells mature normally since they can interact with cells in the intestinal lymphoid tissues. If there are not enough beneficial bacteria to multiply, the immune system will easily malfunction.

Apart from stress factors, the five As in our lives, namely antacids, antibiotics, alcohol, analgesics (Advil), and animal foods, affect the number of beneficial bacteria. These substances will be accompanied by infection and other drugs to change the ecology of the intestinal microbes, destroy the intestinal barrier and allow food to penetrate the intestinal-related lymph tissues under the intestinal mucosa and into the blood. When this happens, the immune system may treat the food particles in the blood as foreign invaders and produce antibodies to attack them.

Another important role of beneficial bacteria is its aid in the development of killer T cells in the intestine and its help to distinguish the difference between foreign substances (such as infections or bacteria) and the body's tissues. This is why maintaining the probiotics and intestinal mucosa in their optimal state and repairing the intestinal tract is fundamental in maintaining a healthy immune system. A healthy intestine not only helps prevent autoimmune diseases but also treats symptoms of immune deficiency and cures the immune system.

Poison

Poisons refer to any foreign environmental chemicals, heavy metals, or other compounds that produce harmful reactions to the body. Molds are also included because they often produce dangerous poisons. Exposure to environmental toxins may damage the immune system and other cells in the body and cause autoimmune diseases. We are poisoned to an unprecedented degree in modern society. The U.S. Centers for Disease Control's Fourth National Report on Human Exposure to Environmental Chemicals tested 22 chemical substances and found that the blood and urine of most Americans contained all of the above chemical substances. This is not surprising as we often come into contact with poisons through food, pesticides, groundwater, industrial waste, and industrial chemicals.

For autoimmune diseases, academia is particularly concerned about any poisons that may alter deoxyribonucleic acid (DNA), as well as ribonucleic acid (RNA) and the chemical structure of cellular proteins which carry genetic information. This is because these poisons change the tissue structure of the human body, causing the body to attack its tissues as foreign objects.

The most researched toxicant related to autoimmune diseases is mercury (the 22 toxicants reported by the Centers for Disease Control, ranked sixth), and mercury exposure comes from silver powder amalgam that fills tooth decay. It is also the product of burning coal or wood and incinerating mercury-containing substances. Mercury in the air has entered the soil, rivers, and oceans, and is present in fish that people eat, such as sailfish, tuna, silver bass and king mackerel (the higher up the food chain, the higher the mercury concentration. Large fish, which consume small fish, tend to contain the highest mercury content).

Studies have confirmed that mercury is related to Hashimoto's thyroiditis, Graves' disease, lupus erythematosus, and multiple sclerosis. It causes the immune system to treat body tissues as foreign objects – as one of the poisons that directly damage tissues.

Another problem with poisons is that when the body is filled with too many poisons, the liver, the body's primary detoxification organ, will overwork as it has to discharge them. This is because the liver has multiple detoxification pathways, and it is the enzyme system responsible for detoxification. Every enzyme system requires specific nutrients. If there is too much poison and insufficient nutrients, the liver will be depleted, causing poison to accumulate. It is also responsible for assisting the processing of naturally produced hormones in the body. If the liver is fatigued due to high levels of poisons in the body, it is difficult for it to process the hormones and chemicals that the body naturally produces. Among them, estrogen is metabolized by the liver. The liver's specific enzyme system needs to function normally for it to process and metabolize estrogen normally. But if the liver is under pressure, estrogen will continue accumulating, causing the body to produce more toxic estrogen, resulting in DNA damage and triggering an immune response. Toxic estrogen is considered an important predisposing factor for lupus erythematosus and rheumatoid arthritis.

3.6 The Battle Between a Tumor and Immunity

The battle between humans and cancer has been longstanding, but to this day, cancer remains a disease that causes fear. Traditional cancer therapies are mainly surgery, radiotherapy, and chemotherapy. However, with the advancement of medical science, treatment therapies such as immunotherapy, targeted therapy, intervention, and radiofrequency continue emerging to provide cancer patients with new treatment approaches. Among them, tumor immunotherapy as a new treatment therapy has become immensely popular in the field of tumor treatment research.

3.6.1 Tumor's Evasion from Immunity

It is precisely because the immune system works around the clock that our health can be effectively protected. When foreign objects (bacteria, viruses, fungi, parasites) invade, it can immediately deploy immune cells with varying functions to manage them. When abnormal cells such as cancerous cells appear in the human body, they can play a monitoring role by detecting and removing them in time to avoid tumors.

Under normal circumstances, immune T cells in the human body can monitor and eliminate tumor cells. However, these cells can also evade the immune system by disguising themselves. Science has discovered a protein called PD-L1 on the surface of many types of cancer cells. When it binds to PD-1 on the surface of immune T cells, the T cells will reduce proliferation or lose their activity, thereby losing their ability to recognize and fight tumor cells and enabling these cells to escape the invasion of the immune system.

Key targets like PD-L1, which can inhibit the function of immune cells, are called immune checkpoints. By inhibiting these targets, drugs, which reactivate the immune function, are known as immune checkpoint inhibitors.

Currently, the common immune checkpoints discovered are PD-1, PD-L1 and CTLA-4. PD-1 is a transmembrane protein mainly expressed in immune cells such as activated CD4+ T cells, CD8+ T cells, B cells, natural killer cells, monocytes, and dendritic cells. Its main function is to promote the maturity of T cells. Under normal circumstances, PD-1 regulates the body's immune response to foreign or self-antigens by regulating the differentiation direction of T cells in peripheral tissues to prevent an excessive immune response.

PD-L1 is a protein that hurts the immune system. It is mainly expressed in antigen-presenting cells, B cells, T cells, epithelial cells, muscle cells, endothelial cells, and tumor cells, and participates in tumor-related immune responses. PD-L1 is expressed very low in normal human tissues, but very high in cancers such as lung cancer, colorectal cancer, and ovarian cancer. As it is mainly expressed in tumor cells, immunotherapy methods using PD-L1 antibodies to kill tumor cells are currently being extensively studied.

CTLA-4 is a transmembrane receptor on T cells. CTLA-4 binds to B7 to induce T cell anergy and participates in the negative regulation of immune response. In 1996, James P. Allison's research group proved that the use of CTLA-4 antibodies can enhance immune function, thereby inhibiting the occurrence and development of tumors.

Corresponding to immune checkpoints are immune checkpoint inhibitors, which can block their immunosuppressive effects and activate the immune system, including Nivolumab, Pembrolizumab, and Ipilimumab.

Nivolumab is a fully-humanized monoclonal antibody. By blocking the binding of PD-1 to its ligands PD-L1 or PD-L2, it reverses the state of tumor immune escape and restores the activity of T cells to kill tumors, thus achieving its objective of suppressing tumor growth. It is the first anti-PD-1 antibody drug to enter phase I clinical trials. It is currently widely used and has shown good therapeutic effects in the treatment of various malignant tumors.

Pembrolizumab, a humanized monoclonal antibody that inhibits PD-1, is one of the drugs approved by the US FDA for the treatment of advanced melanoma. It has a high affinity with PD-1 and almost eliminates immunogenicity and toxic side effects.

Ipilimumab, a fully human monoclonal antibody against CTLA-4, was first used in the treatment of melanoma. Because of its good curative effect, it has been approved by the FDA for the treatment of advanced melanoma in March 2011. However, due to the limited efficacy of Ipilimuma on its own, it is now used in combination with other treatment options.

Simply put, the role of immune checkpoint inhibitors is to inhibit their combination through different ways to achieve the function of reactivating T cells, allowing it to restore its normal function of recognizing tumors and controlling tumor progression. Therefore, they act as a brake system that can slow down the activity of these organelles.

3.6.2 Regulation of Tumor Immunity

The interaction and influence between tumors and immunity is a huge and complex mechanism. The ability of tumors to destroy immunity is not limited to the tumor microenvironment itself as it can also cause extensive and variable damage to the body's immune system.

In terms of regulating immune cells by the tumor microenvironment, it includes the regulation of immune cells derived from lymphoid and myeloid lineage.

Among the lymphoid line-derived immune cells, CD8+ T cells and natural killer cells can effectively kill tumor cells. However, in the tumor microenvironment, the functions of these two types of cells are inhibited to varying degrees. Tumor-infiltrated CD8+ T cells often develop T cell tolerance due to differentiation failure, cell failure, and other reasons, and cannot recognize and kill tumor cells. The main inhibitory receptors of natural killer cells include PD-1 and NKG2A. In addition, studies have found that natural killer cells have other immune checkpoints. For example, high expression of IL-1R8 in natural killer cells can inhibit the maturation of natural killer cells, and tumor cells can inhibit the activity of natural killer cells by producing prostaglandins.

Myeloid-derived immune cells primarily include tumor-associated macrophages and myeloid-derived inhibitory cells, which are an important part of the tumor microenvironment to exert immunosuppressive effects. They play their role of promoting tumors by primarily promoting the growth and proliferation of tumor cells and inhibiting the function of T cells.

In addition, apart from the interaction between immune cells, they can regulate mesenchymal cells and endothelial cells, building a cellular regulatory network for the tumor microenvironment. On the one hand, immune cells participate in the formation and regulation of tumor blood vessels by acting on endothelial cells, especially immune cells derived from myeloid lineage. For example, tumor-associated macrophages secrete a variety of cytokines to promote the growth and maintenance of blood vessels in the tumor microenvironment, thereby promoting tumor development and metastasis.

On the other hand, immune cells derived mainly from myeloid lineage secrete a large number of cytokines to recruit and activate mesenchymal cells in the tumor microenvironment. For example, tumor-associated macrophages secrete EGF and platelet-derived growth factor-β to regulate the differentiation direction of mesenchymal cells, as well as TGF-β to activate fibroblasts and stimulate them to produce a large number of collagen fibers. The TGF-β-induced Tregs cells also play an important role in regulating mesenchymal cells.

In terms of the tumor progression's impact on the systemic immune system, the University of California San Francisco research team once published the results of the tumor's systemic immune suppression research in Nature Medicine.

They used mass spectrometry and flow cytometry to dynamically and systematically analyze immune cell subsets in mouse tumors of eight cancer models. It was found that immune components of the tumor microenvironment were significantly different between these models. In general, tumor-associated macrophages account for a very large proportion of various tumors. However, the proportions of tumors vary greatly.

For example, MC38 colorectal cancer and SB28 glioblastoma models have relatively few adaptive immune cells, LMP pancreatic cancer and genetically induced Braf/Pten melanoma models have extensive eosinophil infiltration, B16 melanoma and three breast cancer models (4T1, AT3, and autologous MMTV-PyMT) showed characteristics of relatively low local immune cell abundance with high diversity. This also means that the immune status of tumors varies.

In addition, studies have found that this systemic immune change exists regardless of whether the tumor has metastasized or not, and is also closely related to the primary tumor size in the MMTV-PyMT breast cancer model. Overall, 78.4% of the tumor's remodeling of systemic immunity can be explained by the size of the tumor, and the remaining part is related to lung and lymph node metastasis.

As for the spleen, the remodeling of breast cancer's immunity to the spleen leads to an increase in the frequency of neutrophils, eosinophils, and monocytes, and a decrease in B cells and T cells. In addition, the researchers found that this systemic immune suppression can be reversed by surgically removing the tumor.

4

Immunity and Battling the Pandemic

4.1 The Basics and Implementation of Herd Immunity

During the pandemic, we witnessed different anti-pandemic strategies in different countries. From Italy's blockade of its entire territory to France's announcement of school closures, as well as China's strict controls, all countries used the suppression strategy without social distancing. However, United Kingdom proposed a strategy of herd immunity, which was to let 60% of its population be infected with the new coronavirus, which has a mortality rate of 2%. This was done so that more could develop immunity to the virus, thereby reducing the spread.

For this reason, the British Prime Minister even warned the public to be prepared to lose loved ones. It received a response akin to a stone raising a thousand ripples. The editor-in-chief of The Lancet angrily criticized the British government for failing to follow science in pandemic prevention and ignoring important evidence from China and Italy, saying that "the government is gambling with the people's lives." So, what exactly is herd immunity?

4.1.1 Embarking on Herd Immunity

Herd immunity, also known as community immunity, occurs when enough people are immune to the pathogen causing the disease, resulting in protection for those without immunity. The theory of herd immunity indicates that when a large number of individuals in a group are immune to a certain infectious disease or there are few susceptible individuals, the chain of infection that spreads between the infectious disease among individuals will be interrupted.

To produce herd immunity, people must develop immunity after infection. Many pathogens are like this, which means that the infected person will not get the disease again after they recover because their immune system produces antibodies that can defeat the disease.

The calculation of the herd immunity threshold depends on the basic number of infections of the virus R0, which is the average number of infections per patient. Scientists estimate that the R0 value of the new coronavirus is between 2 and 2.5, which means that in the absence of prevention and control measures, each infected person will infect about two people on average.

If we want to know how herd immunity works, we must think about the increase in the number of new coronavirus cases among susceptible people – 1, 2, 4, 8, 16, and so on. But if half of the people are immune, then half of the people will not be infected, effectively reducing the speed of transmission by half. After that, according to the Science Media Center, the spread of the epidemic will become – 1, 1, 1, 1 . . . Once the infection rate is lower than 1, the epidemic will be eliminated.

This also explains the logic behind the strategy adopted by the United Kingdom, which is to cease the prevention and control of the pandemic and allow a large number of people to heal and gain immunity after an infection. In other words, the efforts are spent on preventing "deaths" instead of preventing "infections." While managing the pandemic, we will not sacrifice social vitality and economic development as a result of strict control measures and will strive to minimize the cost of fighting it.

However, while such a "herd immunity" approach seems to be feasible theoretically, there remain many uncertainties on whether it will be successful in reality.

Herd immunity is usually obtained by vaccination, such as the development and the use of the smallpox vaccine, which results in us developing immunity and eliminating the infectious disease. Otherwise, it comes from people who have been in general contact with those who have been infected with the virus. The series of vaccines that children receive after birth, including BCG, hepatitis B, white lily, meningitis, etc, as well as the annual influenza vaccine for adults, follow the logic of reducing the proportion of susceptible populations to curb the large-scale spread of infectious diseases.

Unfortunately, there is no vaccine for the new coronavirus to achieve strong herd immunity over a long period after its outbreak. The fatality rate and the probability of adverse reactions of a correctly inactivated vaccine are almost negligible, but the estimated probability of a healthy person dying from new coronary pneumonia is currently 1–2% and it is higher among the elderly or patients with other diseases. Initially, the British authorities adopted a herd immunity strategy to deal with it, which means that 60% × 1% = 0.6% of the total population will die prematurely. The British government's approach was sliding into a humanitarian crisis, resulting in its controversial policy at the beginning of the pandemic.

In reality, the strategy of herd immunity is not just a scientific issue as it may involve some human and ethical issues, as well as major hidden dangers. With the development of scientific and technological civilization in modern society, people hope that more people can have the right to a better life, instead of being subjected to the "survival of the fittest," which is the law of nature. When people are faced with an infectious disease with a certain proportion of deaths, given the possibility of scientific prevention and control, choosing the seemingly fair strategy of herd immunity may have ethical risks.

The Clinical Ethics recommendation issued by the Italian Society of Anesthesiology and Intensive Care also suggests that medical personnel should consider "longer life expectancy" as a priority factor first and not necessarily follow the "first come, first served" principle in their evaluation. But this measure should be implemented as the final plan when medical resources are severely lacking and only after all relevant parties have made every effort to increase the available resources (in this case, ICU resources).

In addition, the UK's herd immunity strategy is based on the fact that most people are asymptomatic or have only mild symptoms after being cryptically infected by the virus, to obtain universal immunity among the population. However, this strategy is risky for individuals. Some mild patients will suddenly progress to a critical state without warning, and it is extremely difficult to treat critically ill patients. In reality, from the early experience of Wuhan and the development of the epidemic in Italy and Iran, the biggest risk is the uncontrolled epidemic. Once it spreads rapidly and severely ill patients increase, if the hospital does not respond adequately, it may cause a run on medical resources.

4.1.2 The Social Attributes of R0

Achieving herd immunity is closely related to a parameter, R0. R0 is a way for epidemiologists to track the infectiousness of a disease. It is an indicator of the infectivity of a pathogen, showing the average number of people who will spread the disease to others in an environment where the pathogen is not subjected to any external intervention and everyone has no immunity.

The estimated value of R0 is directly related to the spread of infectious diseases. When R0 is greater than 1, it indicates an outbreak and rapid spread of the disease. The value will continue to increase if the spread is not contained. When R0 is equal to 1, it indicates that the disease is endemic and controlled and that it has co-existed with humans for a long time. When R0 is less than 1, it means that the disease is no longer infectious and will gradually disappear.

A paper published in *The New England Journal of Medicine* on 29 January 2020 analyzed the first 425 patients diagnosed with COVID-19 in Wuhan, China. Through the analysis of five groups of human-to-human transmission cases, it estimated that the median time of human-to-human transmission (from the first person to the second person) is 7.5±3.4 days, and the estimated R0 value is 2.2, that is, one patient can infect 2.2 patients.

At the beginning of February 2020, medRxiv uploaded the largest clinical retrospective study of the COVID-19 virus. The study analyzed 8,866 cases as of 26 January, including 4,021 confirmed cases and 4845 suspected patients. The R0 was obtained through the study is about 3.77.

In a report in Emerging Infectious Diseases, a publication by the US Centers for Disease Control and Prevention, researchers updated the R0 value using two modeling methods (arrival model and case count model). Based on the assumption that the consecutive interval is 6–9 days, and the median R0 of the COVID-19 virus was 5.7 (95% CI 3.8–8.9). In other words, a COVID-19 patient can infect 5.7 patients, which is 2–3 times as much as previously thought.

The change in RD reflects how difficult it is to estimate and apply R0. The R0 value includes predictions on human behavior and the biological characteristics of pathogens. It changes in the course of the pandemic, together with factors such as prevention policies and crowd isolation measures. The frequency of person-to-person contact is the parameter that is most susceptible to external factors. It changes dramatically according to factors such as population density, type of social organization, and prevention policies. This is the social attribute of R0.

The accurate calculation of R0 is a very critical part of the prediction of the scale of the pandemic and the formulation of corresponding preventive measures. If the calculated R0 value is lower than the actual value, it will be difficult to curb the pandemic effectively. If the calculated R0 is higher than the actual value, it may result in tougher-than-required preventive measures, incurring unnecessary losses to the economy.

During the pandemic, researchers from Princeton University and the Georgia Institute of Technology analyzed the effects of asymptomatic infections on the prediction of R0 value. The results were published on medRxiv. Researchers used generation intervals (the time from when an individual is infected to when this individual infects others) to showcase the rate of infection in different groups. Through this modeling, they found that if the generation interval of asymptomatic transmission is longer than the generation interval of symptomatic transmission, then R0 will be underestimated. If an infected patient with mild symptoms is not diagnosed and recorded, it will also have similar effects on the R0 values, resulting in a greater systematic error in the estimation of R0.

Based on current experience, we know that the COVID-19 virus is highly contagious and can be transmitted asymptomatically. This increases the transmission rate of the virus and the probability of infection through human contact. To make the correct decisions, large-scale testing is carried out to obtain sufficient data, to avoid a miscalculation in the R0 value among the different groups of people.

Due to the social attributes of R0, it has no fixed and precise value, and will constantly change with environmental factors. In a pandemic, whenever the R0 index is updated, people should also adjust the preventive measures to combat the risks brought by the virus.

4.1.3 Herd Immunity Will Eventually be Naught

We already know that the calculation of herd immunity is completely dependent on the estimation of R0, and because the estimation of R0 is affected by different models and social environments, the value will vary greatly. Apart from the infectiousness of the pathogen, factors such as preventive measures and isolation in different countries and regions must also be considered. These measures will affect the estimated R0 value, which in turn affects the estimation of herd immunity.

Achieving herd immunity requires the close cooperation of the entire society. Taking measles as an example, a widely cited paper discussing herd immunity has provided the estimated R0 and herd immunity for several major infectious diseases. Measles is highly contagious and has a large R0 value. To achieve herd immunity, at least 90% of the population must be vaccinated.

In general, the combined vaccine efficacy and herd immunity threshold required for an infectious disease to disappear is calculated using 1 – 1/R0. When R0 = 2.2, the herd immunity threshold is only 55%. This explains why the British authorities proposed a herd immunity policy in March. The policy aims to let a large number of people contract COVID-19 in the absence of preventive control, for them to heal and gain immunity as they recover, before channeling its medical efforts to treat those who are critically ill.

When R0 is equal to 5.7, the herd immunity threshold rises to 82%, which means that at least 82% of the population must gain immunity against the virus by either vaccination or infection before herd immunity can be achieved, and for the virus to stop its transmission. Italy and Spain, which were severely affected during the pandemic outbreak in March, currently have only 10–20% of their population infected.

Sweden is the country closest to achieving herd immunity in March 2020. It has a population size of approximately 10 million, slightly smaller than that of Hangzhou City, China. In Hangzhou City, a total of 169 cases and 0 deaths were recorded. In Sweden, 9,141 cases and 793 deaths were confirmed. The result of nearly achieving herd immunity is based on the fact that Sweden does not test patients with mild COVID-19 symptoms and that the epidemic curve was still creeping up. Even the President of the USA said that Sweden was trying to achieve herd immunity and was suffering greatly.

While professionals from around the world are working vigorously to develop a COVID-19 vaccine, there will not be a vaccine strong enough to achieve herd immunity against COVID-19 for a long time to come. The fatality rate and the probability of adverse reaction of a correct inactivated vaccine may be negligible. The probability of a healthy person dying from COVID-19 is approximately 1–2%, though this number may be higher for older patients and those with other ailments. If the policy of achieving herd immunity is pursued, 82% × 1% = 0.82% of the population will die prematurely, and the change in R0 value will make herd immunity go to naught.

On 10 April, Zhong Nanshan, academician of the Chinese Academy of Engineering, and epidemic experts from Korea had an online exchange on the preventive measures for COVID-19. Zhong Nanshan mentioned that it is now time to consider developing a vaccine as quickly as possible. Vaccines are man-made, and the method of achieving herd immunity through mass infection in the majority of the population will not work as the sacrifice would be too huge. Human intervention is therefore required.

4.2 The Birth of a Vaccine

The purpose of vaccines is to provide immunity.

A vaccine is a modified virus or part of a virus. When the human body is vaccinated, either orally or by injection, it will produce protective antibodies and immune memory, giving the human immunity against the virus. This is somewhat similar to a military exercise, where humans gain immunity from the virus without being infected with the virus. When the human body encounters a real virus attack in the future, the strengthened immune system can respond quickly to avoid the virus.

The eradication of smallpox was the first time in human history when an infectious disease was completely eradicated by vaccination. Variolation of smallpox was invented by China in the 16th century. In the 18th century, Britain invented the variolation of cowpox. The modern smallpox vaccine was born in the 20th century. In 1977, the last smallpox patient in the world was cured, and this highly infectious disease that had endangered mankind for centuries was completely eradicated.

Vaccines are the protective shield for humans against virus invasion. Infectious diseases which once ravaged the world, measles, polio, hepatitis B, and more were effectively controlled through vaccines. Traditional vaccines include inactivated vaccines, live attenuated vaccines, and subunit vaccines made with certain components of natural microorganisms. New vaccines mainly refer to vaccines developed by genetic engineering technology, such as genetic engineering vector vaccines, nucleic acid vaccines, gene-deleted live vaccines, and others.

4.2.1 The Process of Developing a Vaccine

The first step in vaccine production is to produce an antigen, the most important ingredient of the vaccine. Biologically active substances that can usually be used as antigens include inactivated viruses or bacteria, attenuated strains of live viruses or bacteria, virus or bacterial extracts, effective protein components, toxoids, bacterial polysaccharides, polypeptides, and nucleic acids, which are used in the development of DNA vaccines in recent years.

After the antigen is produced, pre-clinical research is required. This is to ensure that virus strains and bacterial cells are safe, effective, and can be produced continuously. Taking virus vaccines as an example, strain screening is the first step of the pre-clinical research, followed by necessary strain attenuation, adaptation to the cultured cell-matrix, and studies on the strain's multiplication. The vaccine is then checked for quality, and animals for trials are selected. The nature of the vaccine will determine the animals used for experiments, whether it is mice, guinea pigs, rabbits, or monkeys. When safe and effective vaccines with consistent quality can be produced, consistent, application for clinical trials can then be made to the respective state regulatory agency.

Clinical trials are usually divided into three phases.

The safety of the vaccine on humans is usually tested in the first phase, and the testing size can range from tens to hundreds. The second phase focuses on the study of vaccine dosage and the evaluation of the results from the first phase. The testing size is then expanded upon confirming that the vaccine is safe for humans, and the size ranges from hundreds to thousands. The third phase of clinical trials adopts a randomized, blinded, and placebo-controlled (or controlled vaccine) design for a comprehensive evaluation of the effectiveness and safety of the vaccine. Testing size is usually expanded again between thousands to tens of thousands. The third phase of the clinical trials is the prerequisite for vaccines to be registered and approved for sales.

Clinical trials usually require several years, and some even stretch up to more than ten years. Strict safety regulations are in place for each phase of the trial, and there are also strict requirements to meet before each phase can end. Trials for each vaccine may be stopped or even terminated if it is unable to meet its purpose or is unable to fulfill the requirements in place.

Upon obtaining approval for sales and production, companies can produce the vaccine in GMP workshops. The fourth phase of the trial is still required after the vaccine is made available to the public, for studies and long-term evaluation of the safety and effectiveness of the vaccine.

4.2.2 The Five-pronged Approach in the Development of a COVID-19 Vaccine

As the COVID-19 virus is a new virus strain, there will inevitably be many uncertainties when it comes to the research and development of the vaccine. As such, taking a multi-pronged approach would be a sounder solution. So what does each approach entail and what are their pros and cons?

At the beginning of the pandemic, scientific research teams have already prioritized the research and development of a new vaccine. To maximize the success rate, five different approaches were adopted after technical analysis. They were the inactivated vaccine, adenovirus vector vaccine, nucleic acid vaccine, recombinant protein vaccines, and attenuated viral vaccines.

The inactivated vaccine is the most traditional approach for vaccine research and development. It involves in-vitro culturing of the COVID-19 virus and then deactivating it to make it harmless. The inactivated virus is still able to stimulate the body to produce antibodies, allowing immune cells to remember the structure of the virus.

Developing an inactivated vaccine is simple, fast, and relatively safe. It is a commonly used approach to be adopted to manage the spread of acute diseases. Most vaccines available in the market are inactivated vaccines. The Hepatitis B vaccine, polio vaccine, Japanese encephalitis vaccine, and Diphtheria Tetanus Pertussis (DPT) vaccine are examples of inactivated vaccines. Developing a COVID-19 vaccine through inactivation is the most mature and conservative choice. The advantage of this method is that we can ensure that the production and quality are stable.

However, there are also limitations to inactivated vaccines, such as large vaccine doses, short immunization periods, and single immunization pathways. The worst limitation is that these vaccines can cause antibody-dependent enhancement (ADE) and aggravate viral infections. This is a serious adverse effect that may cause vaccine development to fail.

The adenovirus vector vaccine uses a modified and harmless adenovirus as a carrier and is loaded with the S protein gene of the COVID-19 virus to create a vector vaccine that stimulates the body to create antibodies. These vaccines are safe, effective, and cause few adverse reactions. However, there are considerations such as "pre-existing immunity" of the adenovirus that need to be overcome in the development of these recombinant virus vector vaccines. Taking the recombinant COVID-19 virus which is in clinical trials as an example, this vaccine uses Type 5 adenovirus as a carrier. However, most people would have been infected by Type 5 adenovirus when they were growing up, and the body would already have antibodies that can neutralize the adenovirus vector, lowering the effectiveness of the vaccine. In other words, the adenovirus vector vaccine is very safe but lacks in its effectiveness.

Nucleic acid vaccines include mRNA vaccines and DNA vaccines. The vaccine can directly inject mRNA or DNA encoding the S protein into the human body, prompting human cells to synthesize the S protein in the human body and stimulating the body to produce antibodies. Simply put, it is as though a detailed virus file is installed on the body's immune system.

DNA vaccines work by injecting DNA fragments with antigen genes into host cells. These genes are transcribed into mRNA and then translated into antigen protein to induce an immune response. The mRNA vaccines inject specific mRNA sequences of the virus which are synthesized in vitro into human cells, where they are then translated into the protein to stimulate an immune response.

The biggest advantage of nucleic acid vaccines, both DNA and mRNA, is that the virus strain does not need to go through the tedious process of screening, isolation, culture, and passaging. The viral gene sequence can be used directly for the production of the vaccine. It has a low cost and can be exponentially cultured within a short period, saving a lot of time from pre-clinical research.

The limitation of the nucleic acid vaccine is that the efficacy is not high and would require booster shots. It is also difficult to produce these vaccines in bulk. In addition, DNA vaccines also pose a safety risk from DNA retention, given that DNA fragments carrying antigen genes are injected directly into the host cells.

The mRNA vaccines have more advantages than DNA vaccines. The mRNA vaccines retain the advantage of DNA vaccines that can express intracellular antigens, and overcome the shortcomings of DNA vaccines having low immunogenicity, and the possibility of inducing immunity against the vector. In addition, the vaccines do not pose any risk of integration into the host DNA. However, mRNA vaccines have poor stability and degrade easily, posing a challenge for mass application. As it has a short production cycle and contains intracellular antigens, the mRNA vaccine has an irreplaceable advantage in the development of vaccines for a highly variable virus such as cancer treatment, influenza, and HIV.

A global effort in the development of mRNA vaccine is currently concentrated on BioNTech, CureVacAG, and Moderna. These companies have achieved good clinical progress in the research area of preventive vaccines and tumor vaccines (see table below). Among them, Moderna has proved the safety of its mRNA vaccine in 5 clinical Phase 1 projects. However, the overall effectiveness of the vaccine would still need to be verified as it has only progressed to Phase 1.

Recombinant protein vaccines are also known as genetically engineered recombinant subunit vaccines. It uses genetic engineering to mass-produce the S protein, most likely the antigen of COVID-19 for injection into the human body to stimulate antibody production. It is almost like producing key components of the COVID-19 virus instead of the virus as a whole, and introducing these components to the body's immune system.

Recombinant protein vaccines are safe, effective, and can be produced on a large scale. The most successful genetically engineered recombinant subunit protein is the Hepatitis-B surface antigen vaccine. The key factor to produce a recombinant protein vaccine is finding a good expression system, and this is very difficult. Its antigenicity will be affected by the expression system selected; hence the careful selection is required to select the expression system for the vaccine.

The attenuated viral vaccine uses an attenuated influenza virus vaccine that has been approved in the market as a carrier to carry the S protein of the COVID-19 virus, stimulating the human body to produce antibodies against the two viruses. In other words, this vaccine is a fusion of mildly infectious influenza and the S protein from the COVID-19 virus, providing the body with immunity against two viruses. The overlap of the influenza virus and COVID-19 is of great clinical significance. As the nasal cavity is the most vulnerable area to be infected by the attenuated influenza virus, this vaccine can only be administered through a nasal drip.

The attenuated viral vaccine provides immunity against two viruses in a single dose of vaccine. The frequency of vaccination is low, and it can be easily administered, making it an important type of vaccine. Commonly seen attenuated viral vaccines include live attenuated vaccine for Japanese encephalitis, Hepatitis-A, measles, rubella, and chickenpox. Orally consumed ones include rotavirus live attenuated vaccine. The downside, however, is that the research and development process of an attenuated influenza virus vector vaccine will be very long.

4.3 Vaccine Development, a Long and Arduous Process

The COVID-19 pandemic has triggered unprecedented efforts in the scientific world, from understanding the virus's biology to clinical research and treatment methods. The vaccine has therefore brought hope to many and is under the global limelight.

Scientific research teams, including those from China, have identified five different approaches for vaccine research development. These are namely inactivated vaccine, adenovirus vector vaccine, nucleic acid vaccine, recombinant protein vaccines, and attenuated viral vaccines. However, there are many difficulties faced in the process of vaccine development, making it a long and arduous journey.

4.3.1 How Long Can Vaccine Immunity Last?

In June 2020, a peer-reviewed study on medRxiv showed that the antibodies of recovered patients dropped significantly within a few months after infection. This caught the academics' attention and has generated a lot of discussions.

Between March and June, researchers from King's College London, UK, conducted repeated tests on 96 patients and medical staff from Iraq and the St Thomas NHS Foundation Trust, to test if antibodies to the COVID-19 virus are present in their bodies. After a polymerase chain reaction test, the subjects were confirmed to have been infected with COVID-19.

Researchers have found that the level of antibodies against the COVID-19 virus peaks about three weeks after the patient starts showing symptoms and then drops rapidly after that. Although 60% of subjects had a "powerful" antibody response at the start of the infection, only 17% of the subjects' antibodies retain the same effect at the end of the three-month test period. For patients with more severe cases of infection, the level of antibodies in their bodies is higher and lasts longer. For patients with milder conditions, almost no antibodies can be detected in their bodies at the end of the three-month test period.

Herb Sewell, Emeritus Professor of immunology at the University of Nottingham, UK, and consultant immunologist, said that research at King's College seems to show that antibodies against the COVID-19 virus disappear faster than antibodies against other coronaviruses such as MERS. The immune response towards the MERS has at least lasted for several years.

Sewell believes that if the immune response of a vaccine declines like a natural immune response, people must go for vaccination repeatedly. It is normal for the antibody level to decrease after vaccination. If the human body can produce antibodies at a faster rate in subsequent exposure to the virus, it shows that the vaccine is still effective. Most importantly, the body will not always respond to vaccines the same manner it does to infection.

Concerns about the potential drop in antibody levels may prompt academics to examine vaccines more carefully from the response triggered by T cells, another key part of the immune system.

On 15 July, Stanley Perlman, one of the pioneer researchers of the COVID-19 virus in the United States, reviewed the human body's immune response to various coronaviruses in the journal Immunology. The immune response provides an important reference. The article pointed out that after the human body is infected with SARS/MERS-CoV, it has a very long-lasting T cell memory, and only severely ill patients have an antibody response lastly more than 2 years. In contrast, the antibody response of mild patients and patients infected with the common cold coronavirus is weaker and shorter in duration.

In addition, on July 16, the National University of Singapore School of Medicine published a critical article on the T cell immunity of the COVID-19 virus on Nature, shedding light on how the COVID-19 is spread.

This article pointed out that T cell memory can last for decades. Specific T cell response of 36 patients recovering from COVID-19 was first evaluated in the study. Results show that CD4 and CD8 T cells can recognize the antigenic peptide on the capsid protein. More importantly, the study found that 23 SARS patients who had recovered 17 years ago had retained long-lasting memory T cells for the capsid protein of the COVID-19 virus, and these T cells can interact and respond to the capsid protein of the COVID-19 virus. This shows that we still need to exercise more caution when we research how long the immunity against COVID-19 can last.

4.3.2 ADE – A Potential Roadblock to COVID-19 Vaccine Development

Apart from academics' concern about the possible decline in antibody levels, the antibody-dependent enhancement (ADE) is another potential setback to the development of the COVID-19 vaccine. The ADE has caused the development of the SARS vaccine to fail. Although there are currently as many as 183 projects on developing a COVID-19 vaccine, success is still not guaranteed.

Modern medicine taught us that viral infections begin by adhering to the cell surface, and adhesion is done through the interaction of virus surface proteins with specific receptors and ligand molecules on target cells. Specific antibodies against virus surface proteins can often block this interaction and "neutralize" the virus, causing it to lose its ability to infect cells. However, there are some cases where the antibodies play the opposite role, assisting the virus to enter the target cell and increase the infection rate. This phenomenon is antibody-dependent enhancement.

In simple terms, after a person recovers from a virus infection, the antibodies produced by the human immune system will prevent future secondary infections to achieve immunity. In some cases, such as virus mutation, the human immune system mistakenly believes that the virus has been suppressed and will be completely unguarded against the virus, resulting in the patient being infected with the mutated virus. The symptoms of these infections are worse than those without antibodies and tend to be more infectious. This is known as the antibody-dependent enhancement (ADE) effect.

Reports of ADE have been made for viruses such as the dengue virus, the SARS-CoV, feline coronavirus (FCoV), and other coronaviruses. This is also the case for HIV, measles, yellow fever, respiratory syncytial virus, West Nile virus, and Coxsackievirus. ADE is also one of the major obstacles faced in the development of HIV vaccines.

For viruses with the ADE effect, antibodies produced after vaccination may increase the chances of an infection. In the event where the subject already has an asymptomatic infection, vaccination may cause symptoms to be activated, making the subject ill. In addition, ADE can also cause viral infection at the placental cells, increasing the chances of vertical transmission of the virus from mother to child.

ADE has two main operating mechanisms. The first uses the virus-antibody complex to bind to the cells with FcR on the target cell's surface membrane through the Fc segment of the antibody. This provides the virus with entry into the target cell, making the virus more infectious. The other mechanism is whereby the virus-antibody complex binds to a complement to gain entry into the target cell through the complement.

There are many studies on the ADE effect of SARS-CoV. Given that the COVID-19 virus and SARS share similar characteristics, and have the same target cells and receptors, scientists are cautious with the possibility that COVID-19 will also have an ADE effect. This may make it even more challenging for the development of COVID-19 vaccines.

Stanford University reviewed the ADE effects of vaccines and antibodies in Nature on July 13. The review believes that the possibility of antibody-dependent disease enhancement (ADE) is universal, so more attention should be paid to the development of vaccines and antibody treatments. This is because the basic mechanism of antibody protection may expand the interpretation of immunopathology of viral infections and increase the risks of ADE-related diseases.

Research has shown that there is a lack of understanding of ADE, and that there is a way to test or mark to identify and diagnose ADE. Animal testing cannot accurately predict ADE response of vaccines and antibodies in humans is due to the difference in species. As such, no one can predict when ADE will occur, and we must therefore find ways to prevent and treat ADE before vaccination and antibody treatment.

4.3.3 COVID-19 Mutations will Cause More Uncertainty

Apart from the doubt on how long the immunity of the COVID-19 vaccine can last, and the potential impact of the antibody-dependent enhancement (ADE) effect, mutations of the COVID-19 virus have added uncertainty to the already volatile pandemic.

In June 2020, a study published in the journal Cell pointed out that a certain variant of the virus was spreading based on the shared information through genetic sequences.

Researchers call this mutation "D614G" and noted that this variant has nearly replaced the "D614" variant which was previously spreading in Europe and the United States. D614G is a mutation of amino acid and was first discovered in the spike protein of the COVID-19 virus. Before the mutation, am aspartic acid (D) occupies position 614. After mutation, glycine (G) is in its position instead.

Tracing the history of mutation, scientists noticed that the mutation was not the mainstream variant before March, and only accounted for 10% of the tested sequence globally. As the infection in Europe continues to spread, the figure soared up to 67% in March. At the point where data is collected for this paper, the proportion has reached 78%. The mutated strain did not break out globally but spread in the order of Europe-North America/Oceania-Asia.

Research teams also conducted lab experiments on humans, animals, and other cells in addition to checking more gene sequences. The results showed that the mutated virus strain is more common, spread faster, and more contagious than the previous variant, but did not make the disease more fatal.

In response to the study on the mutation of the COVID-19 virus, Dr. Maria Van Kerkhove, Head of the World Health Organisation (WHO) emerging diseases and zoonosis unit, said that the D614G mutation was actually discovered in February and is technically not a new mutation. At the same time, WHO officials said that studies have shown that 29% of the COVID-19 samples collected are mutated, and the mutated strain has spread to many regions, such as the United Kingdom, the Netherlands, Germany, and the Americas.

Although there is currently no evidence that this mutation will worsen the ongoing pandemic, it also brings new challenges to the already uncertain vaccine development. The mutation may also add uncertainties to the already out-of-control pandemic. Will this mutated virus undergo another mutation? What impact will another mutation of the virus have on research?

4.4 The Deathly Accomplice of COVID-19

As we all know, COVID-19 is a "self-limiting disease." Just like the SARS virus from seventeen years ago, there is no targeted medicine for it, and it relies on the body's immune system to heal itself. Of course, there are modern drugs that can interfere with the reproduction of the virus and prevent it from spreading. The combination of the body's immune system and modern drugs will increase the possibility of a full recovery from the COVID-19 virus. However, this is not the case in reality and it has still had a huge cost for people's loves. Why is this so?

4.4.1 The Deadly Accomplice Originates from the Immune System

The deadly accomplice of the COVID-19 comes from the human immune system.

Known human coronaviruses (hCoV) can be divided into low pathogenic coronaviruses and high pathogenic coronaviruses. A low pathogenic coronavirus infects the upper respiratory tract, causing mild, cold-like respiratory diseases. In contrast, highly pathogenic coronaviruses such as severe acute respiratory syndrome (SARS virus) and the Middle East Respiratory Syndrome (MERS-CoV), mainly infect the lower respiratory tract and cause fatal pneumonia.

Severe pneumonia caused by these pathogens is usually associated with rapid viral replication, a large number of inflammatory cell infiltration, and increased pro-inflammatory cytokine/chemokine response, leading to acute lung injury and acute respiratory distress syndrome (ARDS).

During the initial stage rapid virus replication, fever and dry cough are some of the symptoms that will show. Subsequently, the fever will persist and there will be a decrease in blood oxygen levels before pneumonia would develop. The titer of the virus will decrease towards the end, and 20% of patients will develop acute respiratory distress syndrome, eventually resulting in death.

More scientists are starting to believe that the continued reduction of the virus in the later stage is caused by the patient's immune system. The cytokine storm in the patient is so strong that the nucleic acid test results of the throat swab or nasopharyngeal swab become negative, so the patient's lungs enter self-destruct mode which is not induced by the virus. The excessive force of the immune cells eventually became the deadly accomplice of the virus.

4.4.2 What Causes the Cytokine Storm?

To know what causes a cytokine storm, we must first understand what cytokine is.

Cytokines mainly include interferon (IFN), interleukin (LL), chemokines, and tumor necrosis factor (TNF). These cytokines are secreted by certain immune cells, with some of their functions are promoting inflammation while some inhibit inflammation. In a normal human body, this is maintained in a balanced state. Pro-inflammatory cytokines can activate and recruit other immune cells, to secrete more cytokines, activate and recruit more immune cells, forming a positive feedback loop in the process.

When the immune system is over-activated due to infections, drugs, autoimmune diseases, and other factors, it may secrete a large number of pro-inflammatory factors to cause a positive feedback loop. This will eventually break a certain threshold and get out of control, over-amplifying to form a cytokine storm eventually. As a result, immune cells break out from the diseased body part and begin to attack healthy tissues, swallowing red and white blood cells, and even destroying the liver.

Blood vessels will expand to allow more immune cells to enter the surrounding tissues. When the infiltration becomes too severe, the lungs accumulate fluid, and blood pressure drops. Blood clots will appear in various parts of the body, further restricting blood flow. When the organs do not get enough blood supply, the patient may get into a state of shock, causing permanent organ damage and even death.

Randy Cron, a pediatric rheumatologist, and immunologist at the University of Alabama and author of Cytokine Storm Syndrome, stated that most patients who experience a storm will have a fever, and about half will have neurological symptoms such as headaches, convulsions, and even coma.

Early signs of a cytokine storm in COVID-19 patients were first discovered in Wuhan, China. An analysis of 29 patients by Wuhan doctors found that the cytokine IL-2R and IL-6 indicators were higher in severe infections.

Doctors in Guangdong, China researched 11 cases and found that the presence of IL-6 is a precursor of symptoms similar to cytokine storm. Another medical team analyzed 150 cases in Wuhan, China, and found that a series of molecular indicators of the cytokine storm, including IL-6, C-reactive protein, and ferritin, were higher than those in the surviving patients.

How does the cytokine storm occur in COVID-19 patients?

According to research, when a person is infected with the COVID-19 virus, the virus can enter the cell through angiotensin-converting enzyme 2 (ACE2). Therefore, the high expression of ACE2 and lung tissues with direct contact to the external environment have become the main targets of the COVID-19 virus.

When the lung immune cells become over-activated, a large number of inflammatory cytokines will be produced, forming a positive feedback loop that will induce a cytokine storm. A large number of immune cells and tissue fluid then accumulate in the lungs, and these can block the gas exchange between the alveoli and capillaries, leading to acute respiratory distress syndrome.

Once a cytokine storm forms, the immune system will destroy both the virus and normal cells in the lungs, causing severe damage to the lung's function. This is manifested as a large white area on the lung CT, that is, the "white lung." The patient will suffer from respiratory failure, eventually dying from hypoxia.

Organs and areas with high levels of ACE2, such as the vascular endothelial cells, heart, kidney, liver, digestive tract, and other organs may indicate that it is a battlefield between the COVID-19 virus and immune cells. This may eventually lead to multiple organ failure and become life-threatening. Therefore, finding out the key cytokine of the cytokine storm induced by the COVID-19 infection and blocking its transduction will greatly reduce the damage of the inflammatory response to the patient's lung tissue and organs.

4.4.3 Strike Before the Virus Strikes You

When we know that the cytokine storm is beginning or has already begun, we must come up with a way to fight it, to strike before the virus strikes us.

Hormone therapy is usually the first choice to fight the storm. Corticosteroids can enter the cell, bind to the hormone receptor in the cell, and enter the nucleus, to either promote or inhibit the transcription of related genes. Comparing anti-inflammatory drugs to corticosteroids, the latter can provide a full range of anti-inflammatory and anti-cytokine storm effects, and the cytokine storm itself is an all-out attack.

However, corticosteroids also have their limitations. Long-term use of corticosteroids can cause many side effects. Therefore, diseases that require long-term use of hormones, such as rheumatoid barrier and lupus erythematosus, often require the use of drugs as an alternative treatment method. As for the treatment method for COVID-19, C Randy Cron, a pediatric rheumatologist, and immunologist at the University of Alabama said that it is still unclear whether hormone therapy is effective or harmful.

In addition, other drugs can specifically interfere with specific cytokines. If we can describe corticosteroids as the atomic bomb, then these drugs are more like homing missiles which protect their "allies," the positive immune response from being destroyed.

For example, Anakinra, which is a modified version of the IL-1 receptor antagonist protein, has been approved by the U.S. Food and Drug Administration (FDA) for the treatment of rheumatoid arthritis and multi-system inflammatory diseases in children.

Early evidence from China also shows that Tocilizumab may be helpful in the treatment of COVID-19. This drug can block IL-6 receptors and prevent cells from receiving messages from

IL-6. Tocilizumab is commonly used to treat arthritis or to relieve cancer patients undergoing immunotherapy from the effects of the cytokine storm.

In early February, doctors from two hospitals in Anhui, China tried Tocilizumab on 21 severe and critically ill patients. Within a few days, the patient's fever and other symptoms were significantly relieved, and the level of C-reactive protein in most of the patients also dipped. Nineteen people were discharged from the hospital two weeks later. Researchers in the United States, China, Italy, Denmark, and many other countries are continuing to study and test the use of cytokine blockers for the treatment of COVID-19.

5

War Against Coronavirus

5.1 The Pandemic will Pass, but the Lungs will Never be the Same Again

During the COVID-19 outbreak, many people wished the nightmare of 2020 would quickly pass. Others wished for their lives to return to normal after the pandemic. In the months of the coronavirus outbreak, some people have succumbed to the virus, many more were eventually discharged after they were 'fully recovered.'

However, did these people make a full recovery? These recovered patients returned to their lives with one complication or another. If one were to summarise their condition, it would be that they had beaten COVID-19, but at the expense of their lungs.

5.1.1 Typical Pneumonia vs Atypical Pneumonia

Pneumonia can be generally classified into two types: typical and atypical. According to Health News KMUH (A medical publication from Taiwan), typical pneumonia mainly refers to bacterial pneumonia such as the common Streptococcus pneumoniae (pneumococcus). Atypical pneumonia is caused by other pathogens such as the Mycoplasma pneumoniae, Chlamydia pneumonia, Legionella pneumophila, the Rickettsia bacteria, viruses such as SARS, and the new COVID-19 coronavirus.

The characteristics of typical pneumonia include symptoms showing up quickly, them being observable, and acute. These symptoms include high fever, chills, cough, thick sputum, headache, and chest pain. There may also be rust-colored phlegm, heavy breathing and even breathing difficulties. If it is caused by Streptococcus pneumonia infection, complications usually arise within 48 hours, catching people off guard.

The symptoms of atypical pneumonia, however, are less observable. They include tightness in the chest, muscle aches, and dry cough. In some cases, people can be asymptomatic. As a result, many patients miss the golden treatment period, causing sequelae in the lungs. This is especially true for the elderly or those over 65 years old. With a weakened immunity and imperceptible symptoms after contracting pneumonia, they may brush it off as fatigue.

Viral pneumonia is usually rare because the human body has antibodies to fight many viruses, and its symptoms are usually mild upon infection. However, SARS and COVID-19 are new viruses, and the human body cannot produce antibodies in time to help, resulting in many symptoms arising quickly. These new viruses evaded the human immune system in the course of evolution, allowing them to better survive and develop in the human body. This is also the cause of the global pandemic which we see today.

The SARS virus attacks the lungs and the human immune system, causing pleural effusion and irreversible pulmonary fibrosis. The COVID-19 virus causes phlegm build-up in the lower respiratory tract. Clinical anatomy has shown that lungs are choked with excessive phlegm build-up and mild pulmonary fibrosis.

While pneumonia is common, and bacterial pneumonia does not usually leave any sequelae, viral pneumonia, on the other hand, is more prone to leaving sequelae of the lungs.

Viral infection is prone to cause interstitial pneumonia, also known as interstitial lung disease (ILD). This is the general term for clinicopathological entities that cause diffuse lung parenchyma, alveolitis, and interstitial fibrosis. Observable symptoms include dyspnoea, chest X-ray showing shadows on the lungs, restrictive respiratory disorders, lowered DLCO, and hypoxemia. In simpler terms, it is the inflammation of deeper lung tissues that may potentially result in pulmonary infiltration, a complication that is equivalent to a severe lung injury. When the lungs heal, scabs will form, resulting in a hardened epidermal layer which will evolve into pulmonary fibrosis and affect the function of the lung.

5.1.2 Pulmonary Infiltration vs Pulmonary Fibrosis

Pneumonia is different from other respiratory infection and it can cause pulmonary infiltration. In fact, when looking at the difference between pneumonia and other respiratory infection, the most frightening one is that pneumonia can cause pulmonary infiltration.

The lungs are made up of many alveoli. Oxygen enters the blood here and this is where the exchange of oxygen takes place. In the event of a bacterial infection, the lungs will be inflamed. The alveolar sacs will be choked with pus, blood, and water, leaving no room for oxygen. This condition is known as pulmonary infiltration.

Another common definition for pulmonary infiltration is the localized inflammation of the lungs with edema along with the lymphatic system. The swelling is caused by the movement of immune cells to the lungs. When pulmonary infiltration occurs, it is usually a sign that the body is unable to eliminate the pathogens causing the infection, and has obstructed the usual lung functions. Patients who suffer from pulmonary infiltration usually experience shortness of breath, light-headedness, and chest pain. In more severe cases, the respiratory tract is also infected, causing cough, sputum, and fever.

The longer the patient suffers from pulmonary infiltration, the more extensive the damage is to the lungs. The infiltration will continue to spread, eventually affecting the entire lung. The lungs may recover if early treatment is administered. If scabs have started to form, irreversible pulmonary fibrosis will occur. Dead cells from this complication will also affect the function of normal cells, causing the condition to spread in the lungs.

Pulmonary fibrosis was a nightmare for many patients during the SARS outbreak in 2003. Not only is the damage to the lungs irreversible, but it also causes patients to become more susceptible to shortness of breath, cough, and fatigue in their daily lives.

So what is pulmonary fibrosis?

Many people would have heard of liver cirrhosis, a condition where the liver is scarred and loses its original function. The same thing happens to the lungs, also commonly known as "Sponge Lungs," where the lungs harden to resemble a dry sponge, becoming thick, coarse and hard. The lungs cannot take in the air nor send air to the entire body, causing patients to suffer from breathing difficulties and oxygen deprivation to the body.

According to research from the Taiwan Society of Pulmonary and Critical Care Medicine, the remaining life expectancy of a patient diagnosed with pulmonary fibrosis is approximately two to five years. However, this is only 0.9 years for someone diagnosed in Taiwan. With the five-year survival rate being lower than that of breast cancer and colorectal cancer, pulmonary fibrosis is deemed a critical illness with a high mortality rate.

When lung tissues are damaged by medical conditions or other factors, fibroblast cells will be activated to repair the damaged area. Scabs will form, and the cell tissues will harden and thicken. The alveoli can no longer absorb oxygen and expel carbon dioxide normally as before, causing lung function to deteriorate.

The damaged and dead cells will never recover. Their presence among the healthy cells will also affect the healthy cells' function. Therefore, once the lung tissues become fibrotic, their function will be affected and will deteriorate continuously, eventually leading to death from respiratory failure.

Paraquat is a type of herbicide that is lethal to humans as it causes rapid pulmonary fibrosis and death upon consumption. In fact, a small bottle cap full of herbicide is enough to cause fibrotic scarring of the entire lung and cause immediate respiratory failure.

5.1.3 Pneumonia Induced Pulmonary Fibrosis

Pneumonia causes severe lung inflammation and damage and it can easily trigger pulmonary fibrosis easily. The extent of the damage is not as severe for common pneumonia caused by

Influenza A and B virus, but this is not the case for the SARS and COVID-19 virus. These viruses primarily attack and cause irreversible damage to the lungs, and they are feared by many.

A study was conducted to compare the CT images of over 60 patients when they were admitted and before they were discharged. According to the standards set by Novel Coronavirus Pneumonia Diagnosis and Treatment Plan (7th Edition), the odds of patients suffering from pulmonary fibrosis upon recovery is 70%. For severe cases, it is 100% of the time. Additionally, 80% of the patients still suffer from shortness of breath upon discharge.

CT images that were taken during the acute phase of COVID-19 infection usually showed ground-glass opacities and patchy consolidations, with some lesions similar to that of pneumonia. For severe cases, diffuse ground-glass opacities could be observed. Studies of lung specimens and autopsies have also shown diffuse alveolar damage. During the convalescence period, mesh shadow in CT images and traction bronchiectasis, a type of fibrotic change in the lung tissues, were observed. In some cases, these changes resembled that of normal interstitial pneumonia. In other cases, patchy consolidations were seen along with the bronchial vascular bundle without affecting the pleura (non-specific interstitial pneumonia-like changes). The more extensive the area of consolidation is on the lobes, the more the lung functions are affected, posing an uphill challenge for a patient's recovery.

Princess Margaret Hospital, Hong Kong, has also conducted studies on 12 patients and found that two to three of them were unable to make a "good as new" recovery. These patients easily get out of breath when they increase their walking speed, and some of them have a decreased lung capacity of 20–30%. CT scans done for 9 COVID-19 patients yielded results of the images of the lungs resembling that of ground glass, thick, coarse, and foggy, indicating that their lungs are damaged.

COVID-19 is a new disease. Although it is under control, we still have very little knowledge of its characteristics, especially in terms of the pathophysiological effects of post-inflammatory pulmonary fibrosis. Will the post-inflammatory pulmonary fibrosis heal on its own or will it persist? Will the lung function continue to deteriorate? These are questions that can only be answered from follow-ups after patients are discharged.

The only experience that can be tapped into now is SARS, where patients continue to suffer from post-inflammatory pulmonary fibrosis and deterioration of lung functions even after one year upon discharge. Precautions need to be taken and we have to increase follow-up consultations and treatments with recovered COVID-19 patients. This is to ensure that early interventions for post-inflammatory pulmonary fibrosis can be made if required, and for prospective medical drug research. It has been 17 years since the SARS outbreak occurred in 2003, and researchers have gained a deeper understanding of the pathogenesis of pulmonary fibrosis. Evidence-based medicine is available for drug treatment for the different causes of pulmonary fibrosis, and these drugs are approved and available for use in China. Whether COVID-19 patients suffering from post-inflammatory pulmonary fibrosis can benefit from these drugs will be something that we will need to explore.

In addition, adjuvant therapy in the form of muscle training can also be administered to improve muscle endurance and lighten the burden of the lungs to prevent them from fatigue and aggravated conditions. At times, wheezing, coughing and fatigue may be due to weak lung muscles instead of decreased lung function. The muscles could be simply too weak to function properly and require more effort to supply oxygen to the body. Therefore, an adequate amount of muscle training can also help maintain lung function.

It is fortunate that the majority of the COVID-19 cases are mild, and that they have mostly recovered upon discharge. People believe that things will get better with time, and that recovered patients do not have to worry too much. These patients can swim more regularly shortly as a form of recuperation, as the water pressure can help improve lung functions.

5.2 What is the Primal Form of the Virus?

Since the start of the pandemic, the source of the novel coronavirus has been the biggest question. While it is almost certain that it originates from bats, there is still no clear answer of how it was transmitted to humans.

5.2.1 Virus Transmission Blueprint

The report published by Yong-Zhen Zhang and his team from Fudan University in the scientific journal Cell outlines the origin of the viral genome research of COVID-19. This report complements previous articles on the origin of the viral genome and provides a relatively complete blueprint for the spread of the virus, from convergent evolution to cross-species transmission, and how it caused a global pandemic through insidious transmission.

In general, asymptomatic transmission and pre-symptomatic transmission make the novel coronavirus highly contagious. Since it is no longer possible to sample the animals in the Huanan Seafood Wholesale Market and the very first infected person, Patient 0, cannot be identified, it is impossible to find out if the pandemic is caused by cross-species transmission between animals or asymptomatic transmission between humans.

The first case of COVID-19 case reported internationally was a patient admitted to the Central Hospital of Wuhan on 26 December 2019. Zhang and his team received the patient's viral genome sequence on 5 January 2020 and uploaded it onto Genbank. With the information obtained from this sequence, the research team did a comparison with other viruses and found that the virus belongs to the same virus family with SARS and shares 78% genomic homology with it.

The virus which is the most genetically similar to the COVID-19 virus is the RaTG13 virus found in Yunnan, with 96% genomic homology. The most prominent difference between the two viruses lies in the furin cleavage site between the S1 and S2 subunits of the S protein. Another observation was that only one of the six key amino acids that RBD binds to ACE2 in the two viruses was identical. One other important piece of information from relevant reviews shows a unique PAA amino acid sequence found at the S1/S2 site. This PAA sequence was found in another coronavirus, RmYN02, which was found in bat samples from Yunnan in 2019. However, RmYN02 only shares 72% genomic homology with the COVID-19 virus.

There is ample evidence to prove that the COVID-19 virus originated from bats. However, it remains unclear how it is transmitted from bats to humans. Both the RaTG13 and RmYN02 virus were discovered in Yunnan, 1,500km away from Wuhan, and it would take the RaTG13 virus 25–65 years to evolve into the COVID-19 virus. This prompted scientists' suspicion of the role of the intermediate host. Civet cats and camels were intermediate hosts when SARS and MERS-CoV virus were transmitted from bats to humans. A thesis published by South China Agricultural University proved that the RBD of the coronavirus found on pangolin share 97% genomic homology as the COVID-19 virus. This confirms the hypothesis of COVID-19's cross-species transmission chain, and made scientists realize that there might be a wide and diverse pool of coronaviruses in wild mammals which have yet to be discovered.

Apart from the cross-species transmission, it is also highly possible that the virus may have been transmitted to humans between November to December 2019 and started to spread between humans. This may explain the prevalence of L-strain COVID-19 virus in Wuhan.

However, there is no clear answer to whether mutation of the furin cleavage site occurred in the intermediate host or during human-to-human transmission. This suggests that the virus has to be continuously monitored for mutations, especially in the current situation of large-scale transmission, we should pay attention to mutations that cause phenotypic changes to the virus.

We might be able to obtain information on the virus transmission if we look into the samples collected from hospital patients before December 2019. COVID-19 has warned us of the danger and extent of cross-species transmission. The curbing of wildlife trade will also be an important issue to be addressed after the pandemic is over.

5.2.2 The ABC Variants

In April 2020, Dr. Peter Foster, a geneticist from the University of Cambridge, and his research team published a report on Proceedings of the National Academy of Sciences of the United States of America (PNAS USA). The report pointed out that there are three mutated variants of the COVID-19 virus, and they are prevalent in America, Asia, and Europe.

The team analyzed the virus genome of the first 160 confirmed patients obtained from the GISAID database. A "genetic family tree" of the virus was mapped out from the study, and the research team categorized the virus into Variants A, B, and C. These 3 variants are spreading in different parts of the world, with Type A as the primal form of the COVID-19 virus.

Studies have shown that Variant A first appeared in Wuhan, China, and is most similar to the virus strain found on bats. Researchers infer that this is the earliest strain of the COVID-19 virus, and it is the main variant in North America and Australia. Infected Americans residing in Wuhan were also tested with Variant A.

The genetic chart shows that Variant B is mutated from Variant A, and is the predominant strain in Wuhan and entire East Asia. However, the mutation of Variant B is slow, and it was never able to spread beyond the East Asia region before the mutation occurred. It is therefore deduced that the environment and host's immunity against the virus outside the region may have made it difficult for Variant B to spread out of East Asia. Researchers call this the Founder Effect, and this is observable in HIV too.

Variant C is mutated from Variant B, and is predominantly spreading in Europe, California, Brazil, Singapore, Hong Kong, and Taiwan. Interestingly, the variant is not found in Wuhan. Another important finding from the study is that the variant was first imported to Italy from Germany on 27 January, and the variant which was spreading in Italy originated from Singapore.

The report published in PNAS USA is not the first report on the evolution and migration of the COVID-19 virus. Peking University's School of Life Sciences published a research report of the COVID-19 virus' L and S variants on NSR earlier in March, and it was widely shared and reported in the media.

A total of 103 viral genome sequences were selected for the study, and findings show that the COVID-19 virus may have come about from natural selection, through convergent evolution, mutations, and viral recombination. Further analysis of the 103 virus genome mutations was conducted, and the study classified these mutated viruses into S and L subtypes, with the S subtype as the ancestral form of the L subtype.

Something worth pondering is that of the 27 virus sequences selected from Wuhan, 26 are L subtype and only 1 is S subtype. In contrast, among the virus sequences selected outside Wuhan (33 domestic, 40 foreign), 45 are L type while 28 were S type.

5.2.3 Tracing the Origin of the Virus

While the pandemic continues to spread globally, the debate on the origin of the virus has also intensified. There is only one scientific method to trace the virus's origin. It is not through political means nor propaganda, but through doing a phylogenetic analysis through virus sequencing.

We are now able to trace many viral pandemics in the past, such as the 1918 Spanish Flu, because Julie Gerberding and the team from the United States Centers for Disease Control and Prevention (CDC) were able to reconstruct the virus genome from the remains of an Alaskan patient in 2005. Philippe Lemey from Oxford University was also able to track how HIV began cross-species transmission in Kinshasa, Democratic Republic of Congo from as early as 1920 through the reconstruction of the HIV genome from the earliest patients in Africa.

All current research on the origin of the COVID-19 virus obtains its virus sequence from two public databases, Genbank and GISAID. The first virus sequence of the COVID-19 virus in both databases is from the first reported COVID-19 patient.

This patient sought medical assistance at Wuhan Central Hospital on 26 December 2019. On 5 January 2021, Yong-Zhen Zhang from Fudan University isolated the virus sequence from the patient and uploaded it onto Genbank, making it the source of all other viral sequences present currently. All phylogenetic analyses conducted will also have the same results of this sequence as the original sequence of the virus. This is why the results of two research studies on the virus origin show why Type A and Type S variants, the original strains first discovered in Wuhan, became more increasingly uncommon in Wuhan over time.

Susan Sontag wrote in her book, *Illness as Metaphor*, "Any important disease whose causality is murky, and for which treatment is ineffectual, tends to be awash in significance." The moral high ground for the debate on the origin of the COVID-19 virus is not about disregarding each other's views, but in clearing the informant's name.

5.3 Demographic Changes during the Pandemic

Throughout human history, wars, famines, and plagues have led to a surge in human mortality. After these disasters, a birth boom occurred. For example, the Spanish Flu killed 25–40 million people in the last century, but the world population quickly recovered during the Baby Boom period.

The COVID-19 pandemic has caused a large number of deaths worldwide and the numbers are still increasing. So what impact does this have on the fertility rate during the pandemic? Under the backdrop of a declining birth rate worldwide, how will the demographic change after the pandemic ends?

5.3.1 Fertility Risks during Lockdown

The first impact which the pandemic has on fertility is the economic loss from the global lockdown. It has also brought about a surge in unemployment, driving up prices of commodities. With a decrease in income, people are less willing to spend, and the uncertainties give rise to increased costs in raising a child. As such, many families have postponed their long-term investments, including having a baby.

Researchers from the Melbourne School of Population and Global Health, University of Melbourne, Australia published a paper titled *The impact of COVID-19 on the reproductive health of people living in Australia: findings from an online survey on medRxiv,* a preprint server for Health Sciences, in August 2020.

In this study, researchers surveyed 518 women below the age of 50 on their intention to get pregnant and the contraceptives they are using. The result was analyzed using descriptive statistics, and they found that the COVID-19 pandemic has affected many participants' plans of getting pregnant. A number of the participants had to postpone their plans to conceive, and some have even decided not to have any children.

The COVID-19 pandemic has also posed a higher risk to pregnant women. As a vulnerable group in society, pregnant women often require more care. But the pandemic is a catastrophe that ravages lives regardless of what group one is in the society.

Statistics released by the United States CDC in June 2020 showed that of the 91,412 pregnant women who were infected with COVID-19, 8,207 were admitted to the intensive care unit (ICU). This is 50% higher than that compared to women who are not pregnant. In addition, the likelihood of requiring ventilators is also up by 70%, with the risk of death remaining as high for them.

A public health agency in Sweden has also conducted research using data of pregnant women and postpartum women COVID-19 patients who received treatment in ICU between 19 March to 20 April 2020. Results show that as compared to a non-pregnant woman of the same age, the proportion of pregnant women and postpartum women who went into ICU was significantly high, by nearly 6 times.

Researchers pointed out that pregnant women may suffer from more severe viral infection partly due to a weakened immune system to avoid an immune response to the fetus. The mother sacrifices her immune defense to protect her baby.

In addition, a study from Weill Cornell Graduate School of Medical Sciences, Cornell University found that the risks faced by pregnant women do not end after childbirth. Researchers did a 4-week study with 675 pregnant women from 3 hospitals in New York from March to April. 70 women were infected with SARS virus-2, of which 9 of them (13%) suffered from at least one of the following three complications: fever, hypoxemia and re-admission to hospital. In contrast, only 27 out of the 605 (4.5%) uninfected women suffered from these problems.

Pregnant women also face the risk of vertical transmission. This is also known as mother-to-child transmission or perinatal transmission, where pathogens are transmitted from the parent to the offspring through the placenta, birth canal, or through breastfeeding. The possibility of mother-to-child transmission has not been confirmed nor ruled out.

A follow-up study on babies born to 33 pregnant women diagnosed with COVID-19 in Wuhan Children's Hospital was published by the Journal of the American Medical Association (JAMA). During the study, researchers found that 3 cases (9%) of the nasal and anal swab test for the COVID-19 virus showed early-onset infection.

In another research published on JAMA, the result shows that vertical transmission is possible, and the clue comes from the IgM antibody.

When a human is infected by the virus, the immune system will produce a large number of immunoglobulins. Among them is the IgM antibody, the first type of antibody produced by the body. This antibody is produced between 5–7 days, has a short maintenance period, and disappears quickly. A positive blood test result indicates early infection. IgG antibody takes 10–15 days to produce, and can remain in the bloodstream for a longer time. A positive blood test is used as an indicator of infection or past infection. For mothers and babies, the difference between the two antibodies is that IgG is the only antibody that can pass through the placenta to provide passive immunity to the fetus, and protect the baby from bacteria and viruses after birth. IgM antibody a larger molecular structure, and do not usually enter the fetus through the placenta.

Researchers from the Zhongnan Hospital of Wuhan University analyzed 6 cases admitted between 16 February to 6 March 2020. The result showed that strict preventive measures were implemented during delivery, and the 6 newborns to mothers infected with COVID-19 had negative RT-PCR results from throat swabs and blood tests. There were also no symptoms of COVID-19. However, blood plasma tests showed a significantly higher level of cytokine, Interleukin 6 (IL-6). Of which, two babies had increased levels of IgG and IgM antibodies, and three babies had increased levels of IgG but normal levels of IgM.

In addition, a report published by a French research team in the scientific journal, Nature Communications showed a newborn confirmed with COVID-19 virus and was infected in the mother's womb. The mother was infected with COVID-19 in her third trimester. When the baby boy was born in March, he started showing severe symptoms 24 hours after birth, including severe body stiffness, restlessness, and other neurological symptoms. Mother-to-child transmission of the virus has not been ruled out, and this may have provided a new piece of evidence on it, dealing a blow to the fertility rate.

5.3.2 Will There be a Post-pandemic Baby Boom?

The impact of the pandemic on the fertility rate during the global lockdown is obvious. Will the global fertility rate continue to decline, or will there be a post-pandemic baby boom like in the past?

A possible reference to answer this question would be the Spanish Flu in 1918. The pandemic caused a 13% drop in the birth rate in the United States between 1918 and 1919. The reasons for the drop included a high mortality rate in adults (and an even higher one for adolescents), pregnant women, and a high stillbirth rate. This caused people to reduce social interactions. It is worth noting that the baby boom in 1920 was not only due to the Spanish Flu pandemic, there was also written work that suggested that it could be closely related to the end of the first World War.

The post-Spanish Flu baby boom may have led to rapid growth in population, but the COVID-19 pandemic is different in many ways. These differences could lead to a different trend in population growth as compared to the Spanish Flu. Researchers from Bocconi University, Italy recently published a report titled *The COVID-19 pandemic and human fertility* in the *Science* journal. The report stated that the COVID-19 pandemic may not bring about another baby boom, but may instead cause the global fertility rate to dip further.

Firstly, the Spanish Flu claimed 50 million lives, with young people at the highest mortality rate. In contrast, the majority of the people infected with COVID-19 are the elderly. The elderly have a weaker immunity system. Apart from being more susceptible to contracting the virus, their conditions are also more likely to aggravate. According to the Malthusian Theory of Population, the death of young adults is more likely to bring about an increase in fertility rate while the death of the elderly has a smaller impact.

Next, the COVID-19 pandemic occurred against the backdrop of a declining global fertility rate. Over the past 50 years, all parts of the world have seen a decrease in fertility rates and this trend is continuing. The global fertility rate was 4.96 in the mid-20th century but has since dropped drastically by 49.22% to 2.52 in 2015.

Changes in the global fertility rate occurred in the 1990s, when the fertility rate of women in some European countries dipped below 1.3, ushering in a new era of extremely low birth rates. Some Asian countries and regions such as Japan, South Korea, Singapore, and Taiwan also joined the ranks in the 21st century.

The effects of negative growth due to low birth rates and an aging population are long-term, complicated, and irreversible unless there is a significant global migration. This means that the COVID-19 pandemic will not bring about a baby boom, but will cause the global fertility rate to decrease further.

However, this trend is not absolute. The effects of the COVID-19 pandemic will vary in different countries due to varying social and economic factors.

In countries with high earning income, the improvement in women's educational level has been the strongest driving force of declining birth rates in recent decades. Childcare centers and kindergartens are common in these countries, as both parents are able to better commit to their work.

However, the COVID-19 pandemic has forced schools to close for a long period, resulting in parents having to spend more time looking after their children. This is a great burden, forcing many couples who intend to have a second child to delay their plans. In addition, women from these countries tend to conceive at a later stage, making assisted reproductive technology an important aspect for couples who plan to have children. The pandemic has undoubtedly affected the operations of relevant medical institutions, too.

More importantly, the pandemic will cause more harm and damage to countries with high earning income. Due to the high cost of raising children, unemployment or a reduction in income will inevitably cause the fertility rate to decrease. This trend is similar to that of the 2008 economic crisis, where the more severe the economic recession is, the greater the decline in fertility rates.

Researchers are interested in finding out if fertility rates will improve as the economy recovers in low-income countries. High fertility rates can be explained using two economic reasons. Firstly, unpaid child labor is of huge benefit to a poor family, though the value of investments in education is lowered in this case. Secondly, high fertility rates provide parents with a form of security for their old age.

However, socio-economic development and the increase in rural-to-urban migration have reduced the rural population by more than half. These changes have impacted the cost of having a child. At the same time, modern contraceptives are becoming more popular these days, and the fertility rate is adversely affected.

The pandemic has forced family planning centers to restrict their activities or even temporarily close. Women are unable to obtain contraceptives from these centers as a result, leading to an increase in unintended pregnancies. This may lead to a short-term increase in fertility rate for low and middle-income countries, similar to the trend observed in West Africa during the Ebola outbreak.

5.4 Are There Other Complications Arising from COVID-19?

Although the majority of the infected patients have managed to recover from the virus, humans have nevertheless paid a huge price for it. In this group of recovered patients, some shocking complications have begun to surface. In addition to damages to the lungs such as pulmonary fibrosis, sequelae of the COVID-19 virus seem to have also spread to one's hearing, the brain, and various parts of the body.

5.4.1 COVID-19 Damages the Brain

In April 2020, a US media claimed that the COVID-19 virus seems to have caused sudden strokes in patients with mild symptoms. These patients are in their 30s and 40s. According to the news report, hospital reports showed a six-fold increase in sudden strokes among young adults within that two weeks. Most of these patients did not have a prior medical history of strokes, and they exhibited only mild symptoms (or no symptoms in the case of 2 patients). In the last 12 months before this, hospitals were treating on average 0.73 adults under the age of 50 for stroke every fortnight.

The Lancet conducted an in-depth study on 125 COVID-19 patients with severe symptoms in the UK, and found that the most common brain complication arising from the COVID-19 virus was a stroke. In the study, 77 out of the 125 patients had a stroke, and most of them were over 60 years old. Most of the patients suffered from ischemic stroke, which is caused by blood clots in the brain.

Apart from stroke, doctors have also followed up on 125 COVID-19 patients in April and found the neurological and mental complications which could arise from the viral infection.

On mental conditions, 39 patients showed signs of confusion or behavioral changes. Additionally, 7 patients developed encephalitis or brain inflammation. Twenty-three patients were diagnosed with psychosis, dementia-like symptoms, and mood disorders. Researchers believe that these diagnoses are new, and they cannot guarantee that some of these patients have not been diagnosed with these symptoms before their hospitalization for COVID-19. Among the 37 patients who displayed changes in their mental state, 18 of them were below 60 years old while the other 19 were older.

Delirium is also considered as another brain complication of the COVID-19 virus. The American Psychiatric Association (APA) defines delirium as an acute disorder of brain function. It is also known as ICU psychosis, ICU syndrome, acute mental disorder, and acute brain failure. Delirium is a common condition in critically ill patients warded in the ICU. Eighty percent of these patients have had such symptoms.

Simply put, patients with delirium will experience a sudden change in their mental state, or will fall into confusion out of the blue. This may last from several hours to several days. Delirium is a short-term illness, but it happens suddenly with strong reactions. It is therefore common to see mentally sound patients having psychotic attacks when their conditions aggravate and when they are sent into the ICU.

The reason why ICU delirium is common among COVID-19 patients is because of the common external environment in which they are in – sedation and hypoxia. COVID-19 patients with severe conditions are usually in a state of hypoxia when they are admitted to the ICU. They would usually require the use of a ventilator over a long period. This means that the longer they need to be on the ventilator, the higher the dosage of sedatives needed, creating a perfect environment for ICU delirium to occur.

Data from earlier research conducted in hospitals showed that one-third of all COVID-19 patients showed signs of delirium. Of which, two-thirds of the critically ill patients showed clear signs of delirium. This high proportion has caused concern among medical staff as these patients would require a longer stay in hospital when they become immobile and weak from delirium attacks. Delirium can also cause more complications, sometimes developing into cognitive impairment such as dementia and other long-term illnesses.

Some studies found many recovered patients suffering from central nervous sequelae.

A neurological study on hospitalized COVID-19 patients was conducted by Huazhong University of Science and Technology, China. The study showed more than 30% of the 214 patients developing neurological symptoms. These symptoms can be grouped into three categories, symptoms affecting the central nervous system such as headache, lightheadedness, impaired consciousness, epilepsy, and acute cerebrovascular diseases; symptoms affecting the peripheral nervous system such as decreased sense of taste and smell, loss of appetite, and neuralgia and musculoskeletal injuries.

According to Japan's NHK report, a 24-year-old man living alone in Yamanashi Prefecture, Japan, developed a high fever after being infected with COVID-19, accompanied by headaches and weakness in his limbs. After being sent to the hospital for emergency treatment, he was diagnosed with meningitis and received a positive cerebrospinal fluid PCR. Experts said that there was a very high likelihood of it being caused by COVID-19. After he woke up, he realized that he had lost his memories of nearly one to two years.

His attending doctor, Moriguchi Takeshi, said that this was an important example of COVID-19 invading the central nervous system. As for why the patient lost his memory, experts believed that this was because the part of the patient's brain responsible for memory was damaged, leading to after-effects such as memory loss. According to the observation results of pathological tissues, many who were infected had pathological changes in the brain such as brain tissue hyperemia, edema, and partial neuronal degeneration.

5.4.2 Auditory Neuropathy

A team from the University of Manchester, United Kingdom conducted an investigation on 121 patients who recovered from COVID-19 eight weeks after they were discharged from the hospital and compiled the results.

Among the 121 recovered patients, 4/5 were mild, and most of them made a full recovery. However, some patients with severe symptoms seemed to suffer from "sequelae" after recovery. Additionally, 16 patients (13.2%) complained of hearing loss, of which eight of them felt that their hearing had deteriorated, while the remaining eight developed tinnitus characterized by hearing ringing, hissing, or buzzing sounds in the absence of an external source.

Researchers added that modern medicine is aware that viruses such as measles, mumps, and meningitis can cause hearing loss. Hence, coronaviruses may also damage nerves transmitting information to and from the brain. He added that COVID-19 may cause problems in parts of the auditory system such as the middle ear or the cochlea.

This means that "acoustic neuropathy" may be one of the sequelae. While the cochlea remains functional, it faces the problem of transmission along the auditory nerve to the brain, making it difficult for the affected person to join conversations with a lot of background noise. For example, it may be more difficult to hear others in a busy and noisy place.

5.4.3 Impaired Male Sexual Functions

In March 2020, the Department of Urology of The Affiliated Suzhou Hospital of Nanjing Medical University researchers uploaded a paper on preprint platform medRxiv (not peer-reviewed), titled *ACE2 Expression in Kidney and Testis May Cause Kidney and Testis Damage After 2019-nCoV Infection*. It suggested that while COVID-19 infection predominantly causes damage to the lungs and immune system, theoretically, it may also cause testicular damage.

According to this paper published on medRxiv, the research team studied three existing clinical data sets, including 6, 41, and 99 patients. The first sample was a family cluster infection, while the other two data sets showed that 3% to 10% of people infected with COVID-19 suffered from abnormal kidney function, manifested by increased creatinine or urea nitrogen. In addition, 7% of patients developed acute kidney injury.

COVID-19 invades cells and causes tissue damage by binding to host cell angiotensin-converting enzyme 2 (ACE2) through spike glycoprotein (S protein). In addition to the lungs, many organs, such as the testis, small intestine, kidney, heart, and thyroid of the human body express ACE2. Among them, the testis expresses a large amount of ACE2 mainly in testicular spermatogonia and support cells and mesenchymal cells, which are closely related to the male reproductive functions.

The seminiferous tubules, which produce sperm, contain various levels of spermatogenic and supporting cells, as well as interstitial cells. Spermatogonia are the precursor cells that form sperm, and supporting cells are essential for maintaining the microenvironment of sperm development. Adult supporting cells no longer divide and are constant in numbers, while mesenchymal cells are the main cells synthesizing and secreting androgens, which can promote spermatogenesis and male reproductive organ development, as well as maintain male secondary sex characteristics and sexual functions.

Therefore, it is theoretically predicted that COVID-19 may cause testicular damage and affect sperm production and androgen synthesis. Obstruction of sperm production can affect male fertility, with a possibility of severe cases leading to male infertility. Androgen deficiency can affect male secondary sex characteristics and sexual function, reducing the quality of life.

The paper concluded that its research confirmed that ACE2 has a high expression in kidney and testicular tissues, which helps to understand the abnormal renal function and mechanism of kidney damage in patients with COVID-19. The results also showed that patients may develop orchitis. Hence, it is necessary to evaluate and follow up on the reproductive function of male patients with the virus, especially young male patients.

Therefore, for men with a history of infections, especially those with fertility needs, it is best to perform fertility-related tests, such as semen quality and hormone levels after recovery, so that problems can be discovered in time and dealt with as soon as possible.

5.4.4 Liver, Kidney, and the Heart

According to Canadian kidney experts, 8%–12% of COVID-19 patients in Canada suffer from severe kidney damage. A nephrologist at Yale School of Medicine in the United States said that 20%–40% of intensive care patients with COVID-19 in the United States suffer from renal failure and require dialysis.

Previous studies have pointed out that kidney damage is common in people infected with COVID-19. A paper titled *Caution on Kidney Dysfunctions of 2019-nCoV Patients* investigated 59 infected patients from Wuhan, Huangshi, and Chongqing. It found that 63% of patients showed symptoms of renal insufficiency, and CT scans showed that 100% of patients had renal imaging abnormalities.

The researchers said that the liver and kidney damage of infected persons may be caused by lung dysfunction. When the respiratory function is impaired, oxygen exchange is blocked, and other organs will be in a state of hypoxia. The kidneys, which are particularly sensitive to oxygen, often suffer severe functional damage.

In addition, a 48-year-old patient from east London revealed that COVID-19 resulted in severe heart damage, while Dr. James Russell, professor of medicine at the University of British Columbia, and several other researchers found that more than half of the critically ill patients suffered heart damage and lost their lives according to a sample survey of 8000 patients.

5.5 What Imprints did the Pandemic Leave in our Hearts?

Takeshi Kitano said, "The disaster is not the death of 20,000 or 80,000 people, but the death of a person happening 20,000 times." Although the COVID-19 pandemic has affected the global population significantly, little is known about its potential impact on mental health. The news about the pandemic is uninterrupted, and everyone is experiencing differing levels of psychological changes every day.

5.5.1 Pandemic Post-traumatic Stress Disorder

The Lancet published the inaugural systematic review and meta-analysis in June 2020, studying the psychiatric consequences of coronavirus infections in more than 3,550 SARS, MERS, and new coronavirus patients who were hospitalized. The results showed that if the course of COVID-19 infection is similar to that of SARS in 2002 and MERS in 2012, most patients who were hospitalized for severe coronavirus infections were able to recover without experiencing mental illness. However, SARS and MERS survivors may still face the risks of mental illnesses such as depression, anxiety, fatigue, and post-traumatic stress disorder (PTSD) for months and years after discharge.

The researchers analyzed that among SARS and MERS survivors, the prevalence of post-traumatic stress disorder 34 months after the acute phase (121/402 cases in 4 studies) was 33% on average, depression after the acute phase was 23 months on average (77/517 in 5 studies), and anxiety disorders one year after the acute phase was approximately 15% on average (42/284 in 3 studies).

As early as the 4th issue of the *Chinese Journal Behavioral Medicine* in 2006, some researchers conducted investigations on post-traumatic stress disorder in SARS patients, front-line medical staff, and the public in epidemic areas. The results showed that the detection rates of PTSD symptoms were 55.1%, 26%, and 31% respectively, and SARS patients had the highest detection rate of PTSD symptoms.

The panic and helplessness arising from the pandemic will gradually subside, and it will eventually be brought under control. Apart from physical health, personal mental health is also particularly important. Therefore, it is worth being more vigilant for post-traumatic stress disorder following the pandemic.

Behind the long-term psychological impact caused by the pandemic is a huge neural activity mechanism. Neuroanatomy studies believe that the brain structures related to mental stress are the prefrontal lobe, amygdala, hippocampus, cingulate gyrus, hindbrain area, and dorsal raphe nucleus, etc. Some researchers pointed out that these changes are often the pathological basis of mental stress-related diseases, especially post-traumatic stress disorder. The prefrontal lobe, amygdala, and hippocampus are particularly noteworthy when studying the biological pathogenesis of mental stress.

In general, the clinical problems of post-traumatic stress disorder patients are mainly memory impairment, which indicates the relation between brain activity and memory. Neuroanatomy researchers have pointed out that the central nervous system may undergo structural plasticity changes related to the size and duration of stress damage, as well as the individual's susceptibility quality under mental stress. In short, strong psychological stress may lead to changes in the structural plasticity of the corresponding areas of the brain.

Studies on animal models of post-traumatic stress disorder also showed that the long-term retention of fear memory and high arousal in post-traumatic stress disorder is closely related to the three brain regions, namely the amygdala, medial prefrontal cortex, and hippocampus.

Among them, the medial prefrontal cortex is related to the de-inhibition of the amygdala and threatening environmental information transmitted by the hippocampus to the amygdala, and its changes may promote the appearance of PTSD symptoms. Increased amygdala activity is the key neural basis for the acquisition, maintenance, and expression of conditioned fear memory. In addition, PTSD may be related to the functional inhibition of the dorsolateral prefrontal cortex, orbitofrontal cortex, anterior cingulate cortex, and excessive activation of the posterior cingulate cortex.

In reality, traumatic psychology is related to the imbalance of functional activities in different brain areas, pathological cognition, and negative emotions. While traumatic psychological experience can be re-evoked, it is different from common memories, which will reappear in the form of strong emotional experience, a physical sensation, or deep visual appearances, such as nightmares or flashbacks. Also, it appears to be permanent and cannot be modified or replaced by future experiences.

Since the memory fragments of PTSD patients are mostly sensory, unconscious, and general sensations related to the experience, one can presume that the recurring and intrusive recurring traumatic experience of such patients may be a reflection of how emotional memory in the cortex (probably the hippocampus) cannot be effectively inhibited and regulated by other cortex (amygdala and prefrontal lobe).

Traumatic psychology may essentially be a person's memories of the painful experience of negative emotional events. The failure to inhibit trauma-related "painful" memories or control the intensity of activating these memories may be due to the important psychological mechanisms, namely post-traumatic stress (PTS) reactions, and PTSD.

5.5.2 The People Affected by the Pandemic

At present, it is believed that the clinical manifestations of post-traumatic stress disorder primarily include (i) recurrent symptoms such as flashbacks (re-experiencing the trauma with heart palpitations or sweating), nightmares and terrible thoughts, (ii) avoidance symptoms such as avoiding locations, objects, thoughts and feelings of the traumatic experience, (iii) awakening and reaction symptoms such as being easily startled, nervous, restlessness or anger and difficulties in falling asleep, as well as (iv) cognitive and emotional symptoms such as difficulties in recalling the key characteristics of a traumatic event, having negative thoughts about yourself or the world and cognitive distortion (such as guilt or self-blame), and losing interest in one's favorite activities.

In addition, sometimes patients experience very severe symptoms, which subside after a few weeks. This is called acute stress disorder (ASD). If the symptom lasts for more than one month and has a serious impact on a person's ability to function, the patient may have PTSD. It is usually accompanied by depression, drug abuse, or one or more anxiety disorders.

Many factors in the pandemic will determine whether a person has PTSD. One part is the "risk factor" that makes people prone to it, and the other part is the "recovery factor" that helps reduce its occurrence. Some of these factors appear before the pandemic, while some may appear during and after.

Hong Kong researchers used the Life Events Self-Rating Checklist and the Hospital Anxiety and Depression Scale to track and study the mental health influencing factors of 131 SARS survivors. Studies found that emotional support (family or friends to talk to) can increase the mental resilience of survivors, while three groups, namely medical staff, people with a history of psychological counseling, and those with others around them, are at a higher risk of displaying psychological symptoms. Another study, conducted 30 months after the end of the pandemic, showed that women and those who had suffered from other chronic diseases were more likely to suffer from long-term PTSD.

On 12 March 2020, the *Journal of Nurses Training* conducted an investigation and analysis of PTSD in frontline nursing staff dealing with COVID-19. The results showed that the PTSD level among them was 40.85±15.81 points, which was much higher than that of ordinary people.

Other risk factors include childhood trauma, minimal social support after experiencing the pandemic, and economic pressure. It can be seen that for ordinary folks, seeking support from friends and family members and finding supportive social groups are the correct measures to help them recover as soon as possible.

The plague is a pandemic, and the people get sick. A large-scale infectious pandemic has never been just a medical event. Fighting one is both a physical and psychological war.

After the pandemic, people experience panic attacks, rebuilding of economies, and resuming of daily lives. The popular searches on social platforms have shifted from the pandemic and returned to pan-entertainment singing and dancing. But, it is hard to deny that the pandemic left indelible harm on too many. Some pain may be gradually forgotten over time, but some may not.

The pandemic reflected the China's progress. The government is becoming more active and efficient, with the media and communication becoming more open and transparent than before. As the world's second-largest economy, the state issued an order to suspend work in all enterprises and ban 1.4 billion people from their homes to control the pandemic. However, the people may have to deal with more challenges thereafter – perhaps the crisis brought about by the economic downturn or PTSD.

According to a research report on the passage of PTSD published by The Lancet, the investigation is aimed at confirming whether PTSD experienced by parents experiencing refugee life is related to a child's mental illness. The study selected 51,793 eligible persons. Among them, 1307 (2.5%) children had contact with mental illness and 7486 (14.5%) child refugees were exposed to their parent's PTSD. From this analysis, it was found that parental PTSD will significantly increase the risk of mental illness in their children.

This also means that PTSD is not entirely an individual affair as sufferers need to manage the cure and recovery from it. From a societal level, we need more care and comfort. We must always remember that 6,000 medical elites flew from various locations in China to Wuhan, frontline doctors spared no efforts in rescuing patients, grassroots members remained at their posts through the storm and ordinary folks helped China in their way to overcome obstacles. At the same time, we should remember the infinite sorrows from the pandemic and every specific person. Perhaps, this may be another form of social therapy for PTSD.

6

Technology Created During the Pandemic

6.1 Revealing the Technology Behind "Leishenshan"

Countless production speeds have accelerated in China during the pandemic. The production of masks increased from more than 43 million to 110 million per day in nine days. Emergency scientific research projects developed kits at an unprecedented speed. Seven projects obtained approvals to embark on clinical trials within a short period. The average time taken by medical teams to receive instructions and assemble to assist Hubei was no more than 2 hours, and no more than 24 hours in Wuhan.

The pandemic has enabled us to witness the unity of the people and the empathy of human nature. Among the countless accelerated speeds in China, both Huoshenshan and Leishenshan Hospitals shocked the world when they were constructed in just 10 days and 13 days. In addition to the requests and hard work of countless people, the technology behind their building works was even more significant – the digital twin technology.

The popularity of digital twins has grown exponentially in the past few years. It is frequently discussed in speeches of major summits and forums and it has attracted attention from inside and outside the industry. So, what exactly is this mysterious-sounding digital twin? What is its connection with the construction of Leishenshan Hospital? What kind of changes will it bring to our lives?

6.1.1 The Past and Present of Digital Twins

A digital twin technology is exactly what its name suggests. According to the internationally recognized definition, it leverages data such as physical models, sensor updates, operating history, and integrates multi-disciplinary, multi-physical, multi-scale, and multi-probability simulation processes to complete the mapping in the virtual space to reflect the corresponding life cycle process of physical equipment.

Simply put, a digital twin is the creation of a digital version of a "clone" based on a device or system. This "digital clone" is created on an information platform and it is virtual. Unlike computer design drawings, the dynamic simulation of physical objects is its biggest feature. In other words, the digital twin is "mobile."

The basis for the "mobility" of the digital twin stems from its physical design model of the physical object, data received from sensors, and historical operations data. The real-time state of the physical object, as well as the external environmental conditions, will be replicated on the "twin body."

In 2002, the concept of digital twins was conceived and proposed by Professor Michael Greaves of the University of Michigan in the United States. He put forth the concept of "virtual digital expression equivalent to physical products" during the product life cycle management course. A digital copy of a specific device or a group of specific devices can abstractly express the real device, as well as refer to it as a basis for testing under real or simulated conditions. The concept stems from the desire to express more clearly the information and data of the device, and hope to piece all the information together for higher-level analysis.

The idea was already put into practice even before the National Space Administration (NASA) Apollo project. In that project, NASA was required to manufacture two identical spacecraft. The one which remained on Earth was called a "twin" and used to reflect (or mirror) the state of the spacecraft performing a mission.

When preparing for flights, spacecraft called "twin bodies" are widely used for training. During the execution of a mission, they are used to perform simulation tests using an accurate space-simulating model on Earth by reflecting and predicting the conditions of the spacecraft and performing the mission as accurately as possible. Thereby they are assisting the astronauts in the space orbit to make the most accurate decision in an emergency. From this perspective, people can also see that the "twin body" is a prototype or model reflecting the actual operating conditions of the object in real-time through simulation.

6.1.2 Revealing the Secret Behind "Leishenshan"

Speaking of this, some people still feel that the digital twin technology is still too far away from our reality, but it is not. Microsoft CEO Satya Nadella described its benefits in New York recently, which he quipped as one of the biggest technology trends.

British retailer Marks & Spencer is increasingly using in-store sensors to create a digital twin of its retail space and the data models to optimize the physical layout of its stores, monitor the temperature of its frozen meat storage, as well as to closely monitor the queues at the checkout stations.

The world-famous Leishenshan Hospital was built using digital twin technology during the pandemic.

The Central South Architectural Design Institute (CSADI) rose to the occasion and designed the second "Xiaotangshan Hospital" in Wuhan-Leishenshan Hospital. The Building Information Modeling (BIM) team of Zhongnan Architectural Design Institute created a digital twin model, established using BIM technology, for Leishenshan Hospital. This technology was used to guide and verify the design, as well as to provide strong engineering support according to project requirements.

A digital twin is a concept that transcends reality and can be perceived as a digital mapping system of one or more, important and interdependent equipment systems.

The Industrial Internet is another such phenomenon. For instance, the installation of machine and production lines, as well as the establishment of a virtual version of the production environment, must be described digitally. This caters to its digitization backdrop.

Using this technology, engineers can not only see the product's external changes but also make it possible to observe the dynamics of internal parts. For example, with the digital 3D model, we can see every change of each component, circuit, and various joints in the engine when the car is moving so that preventive maintenance can be carried out to avoid similar tragedies like the Boeing 737MAX8 crash, which resulted in lives being lost.

As of now, mainstream manufacturers such as Siemens, GE, and Schneider have already begun using this technology to optimize industrial processes.

For a long time, there has been a saying that referred to the "1% revolution in the industrial field," which meant that the production efficiency was increasing by 1% and its costs were reducing by 30 billion. This saying has begun with GE's $245,000 narrative.

In GE's plant, the steam turbine thrust bearing of the power plant changed its position, which changed the axial displacement of the steam turbine from -0.29 mm to -0.445 mm. This level of movement was equal to the width of a single eyelash and was still within the safe operating parameters. However, steam turbines had to withstand extreme temperatures, pressures, and forces. Even small changes may cause huge losses.

However, without intervention, the thrust bearing and axial displacement continued moving. By the time the factory's control center sounded an alarm and discovered the deviation, the damage would have been done. The turbine had to be taken offline, parts needed to be replaced, and electricity production data was lost. The total cost of the electricity generator was $245,000.

GE Digital Industrial Management Service Center engineers in Paris used the "digital twin system" of steam turbines to observe deviations and made advanced predictions to ensure that huge losses like this did not occur.

As of 2018, GE has 1.2 million digital twins, which can handle 300,000 different types of equipment assets. In addition, according to Gartner, "By 2021, half of the large industrial companies will use digital twins, thereby increasing their effectiveness by 10%."

6.1.3 Revolution of New Production Factors is Coming

Advancements in the application of the Industrial Internet of Things (IIot) have brought new vitality to the digital twin. The IIoT extends its value chain and life cycle, highlights its advantages and capabilities based on models, data, and services, and opens up the real path of applications and iterative optimization. It is also becoming an incubator.

As a technology, the urban scene is growing, and the trend to move from industry to urban development is inevitable. It is predicted that by 2022, 85% of IIoT platforms will use some form of digital twin technology for monitoring, and a few cities will take the lead in using digital twin technology to manage smart cities.

From the perspective of domestic practitioners, a digital twin mainly plays a catalytic role in the geographic information industry. For example, drone swarms are used to provide cities with digital models based on image scanning. All functional modules such as streets, communities, entertainment, and commerce will soon have one.

This has been achieved in some current cities. Since December 2018, Shenzhen has started the city's virtual city environment data collection work. It is divided into two phases, namely Phase I and Phase II, which started in December 2018 and April 2019 respectively.

The relevant work is expected to be completed by the end of this year.

Under this model, digitization will be omnipresent at power lines, substations, sewage systems, water supply, and drainage systems, urban emergency systems, Wi-Fi networks, highways, and traffic control systems. This will make urban management much easier.

The digital twin represents the "third wave" of the Internet after search engines and social media. It is the cornerstone of the future physical industry and disruptive technology for product lifecycle management. Whether it is manufacturing, construction, or aerospace, revolutionary changes will occur due to digital twin technology. There is no doubt that it is a revolution in new production factors of modern industry.

6.2 Is the Health Code Really Healthy?

On 11 February 2020, Hangzhou took the lead in launching the health code model, implementing the three-color dynamic management of "green code, red code, and yellow code" for citizens and those who intend to enter Hangzhou. They also linked it with Dingding APP-Workbench. A large number of returning workers applied for health codes via Alipay, and the number of visitors on its launch day reached 10 million.

Subsequently, the health code was promoted nationwide. Leveraging the first local experience, the national integrated government service platform tapped into the technical advantages of Alipay and Alibaba Cloud, and accelerated the development of the national integrated government service platform of pandemic prevention, and control health code system.

Finally, many cities have completed the upgrade of their health codes, realizing the use of electronic health and social security cards, which is to consult a doctor using the code. So, what is the essence of the pandemic prevention health code? Is it an innovation or another function of the health code? Perhaps, a new era is about to begin.

6.2.1 Health is in Essence a Twin

Many people think that when their health code is green, it means that their body is healthy. Of course, there is no issue with this interpretation because this is its basic function to reflect a person's health, especially during a pandemic.

In addition, most people are not the least bothered. The health code allows you to declare your physical condition and apply for it thereafter. This seems to conflict with its basic function. How do you know that you are healthy? How do you prove that you are healthy?

Hence, the problem continues to develop and has reached the stage of exploring the health code's fundamentals, which are to closely integrate people and information to create a digital twin.

Digital twin technology, as its name suggests, represents a "digital twin." Simply put, digital twins are the creation of a digital version of a "clone" based on a device or system. This "digital clone" is created on an information platform and is virtual. The biggest feature of a digital twin is that it is a dynamic simulation of physical objects. In other words, the digital twin is "moving."

When the pandemic just broke out, the government had to set up checkpoints at entrances and exits in various locations, with individuals filling in information registration and information declaration one by one, which was time-consuming and laborious. This was because there was no effective way to efficiently integrate personal identity, travel, and health information.

However, the health code can play a huge role in information exchange. It allows individuals to "carry" their relevant digital information tags. In a way, the health code makes everyone a "digital twin."

If you describe the health code in one sentence, it can be said to be "one person, one code, three-color management." The realization of one code per person depends on obtaining personal historical data information from three dimensions.

The first dimension is space, where the locations visited by an individual can be recorded. The locations are narrowed down to the cities and towns visited, whether one has visited a pandemic area, as well as one's distance from it. The second dimension is time, where the timings at every location are recorded to assess one's time and length of stay at each pandemic area. The third dimension is interpersonal relationships, which includes whether one had close contact with sensitive personnel being one of the factors assessed.

The tracking of individuals from three dimensions enables personal information to be constructed by a three-dimensional twin. While it brings about benefits, it is also accompanied by risks.

6.2.2 Privacy Issues with Health Code

The health code stores data in the cloud so that residents' health information, travel information, and personal identifiable information are bound, stored, and shared within, eliminating the hassle of setting up checkpoints for local authorities and repeatedly filling in health forms. As a digital pass, the health code can be used in the whole city or even across the country. This brings convenience for work resumption and community screening.

Although big data guarantees the smooth flow of information during the pandemic, the realization of information exchange and transparency of personal information also brought about the risk of private information exposure.

When we enter the big data era, among the sea of data, more personal privacy data requires protection or encryption. In the first half of 2019, Internet data breaches surged to more than 3,800, the highest ever. Around 870 million pieces of personal information were sold on the dark web, 773 million email addresses and passwords were stolen and 590 million Chinese resumes were leaked. Apart from the name and phone number, the ID number, household registration, marital status, and home address were all disclosed.

It is estimated that by 2025, 87% of data will require protection. However, the reality is that more than half of it is not adequately protected.

In recent years, incidents of privacy leaks have occurred frequently. We cannot be sure that when the health code reflects a broader definition, the digital twin constructed by the health code can be effectively protected for privacy. Moreover, when the big data personal privacy protection law is seriously lagging, who will be responsible for the troubles and risks caused to individuals in the later stage of this multi-platform, non-standard storage and circulation of personal privacy data?

Hence, we can understand the intense public opposition when Apple announced its joint development with Google on a Bluetooth tool, which anonymously tracks those in contact with new coronavirus patients. Their worries over privacy leaks outweigh their worries over becoming infected. According to a warning issued by the American Civil Liberties Association, any use of mobile phones to track the spread of the new coronavirus requires strict privacy protection.

6.2.3 Automated Administration in the Post-Pandemic Era

With the promotion and popularization of health codes nationwide, will health codes become the new normal in the post-pandemic era? From the initial pandemic prevention to its upgrade, the health code is not just another simple management measure.

You need to know that it is the first" mobile phone-based, 3D facial recognition and multi-occasion population management measure.

Since it is phone-based, it means that it is possible to quickly change control standards at almost no deployment cost. If it is necessary to adjust the control measures for certain groups of people or even a certain individual, it can be done almost immediately.

As it is 3D face recognition-based and is strictly limited to one person per code, it means that it can reach the highest standards of a real-name system. Also, as it is based on backend machine data, it has the strongest anti-counterfeiting features.

Population management on multiple occasions means that the health code can be used in various scenarios. In preparation for a pandemic, it has been gradually upgraded, from being a form of pandemic prevention to serve as an electronic health aid and social security card. It is also gradually used as a form of identification to gain access to specific events.

It can be predicted that the health code will not disappear when the pandemic ends. It will continue to be more widely used for a long time.

When the news reported about a murderer who surrendered after absconding for 24 years without a green code, did we notice that the health code has quietly become a human rights issue? Where is the data management boundary between personal privacy and social supervision? Will the promotion of health codes result in them being tools for the authorities to monitor individuals?

It is clear that with the development of technology, digital twins will become a social trend. A digital automated administration is inevitable, and the continuation of the health code certainly conveys a clear signal that this is the beginning of the administration system encroaching on personal life with unprecedented efficiency and capacity. In such times, balancing is a challenge that people will have to face for a long time.

Of course, the health code provides us with convenient and efficient information exchange during the pandemic, which is also the first step in the popularization of digital twins. This is a node of change and reshaping.

The update of technology means that the boundaries of many industries and scenes have widened and hastened the enthusiasm of promoting digital transformation across all walks of life. In the post-pandemic era, the development of the digital economy will also have a higher strategic position.

While embracing technology, people cannot ignore the accompanying dangers. Will the risks of personal privacy exposure be further aggravated? How does social supervision define the boundaries of data management? With the application and continuous evolution of blockchain technology, can people use technology to solve its problems? Whether the health code is actually beneficial to health depends on how we use it.

6.3 AI Efforts to Combat the Pandemic

With the development and iteration of technology, the concept of artificial intelligence is no longer strange to us. There is a wide range of audio, video, and text information about artificial intelligence in books and on the Internet. Artificial intelligence has rapidly changed from a professional academic term hidden in the laboratory to the mantra of product managers and marketers. It has also become a dinner conversation among the general public.

The changes brought about by artificial intelligence have quickly surfaced. The news you read is based on an algorithm recommendation by artificial intelligence. When you shop online, the homepage reflects products that you are most likely to be interested and buy based on artificial intelligence recommendations. Today, these results have penetrated deep into our work and lives. The technological progress behind these changes in details has become a huge thrust to change our lives.

The outbreak has resulted in huge adjustments to our social lives. It is also an important litmus test for artificial intelligence and a comprehensive digitized test at the national level. With the pandemic, artificial intelligence companies are no longer mere bystanders like before. They play a key role in our daily lives and they improve the overall efficiency of the war against the pandemic.

6.3.1 Virus Diagnosis, Medical Research, and Development

According to WHO, the virus that caused the epidemic, 2019-nCoV, is a positive-stranded single-stranded RNA virus and RNA (ribonucleic acid) coronavirus with an envelope, with a diameter of about 80–120nm.

The important detection method for determining such diseases is through nucleic acid detection. By extracting the nucleic acid sequence in the blood of a suspected case and comparing it with the target virus, the presence or absence of pathogen infection can be fundamentally determined. However, the whole genome sequence analysis and comparison of virus samples from suspected cases is time-consuming and laborious, and artificial intelligence can replace manpower to complete the preliminary screening work, which improves detection efficiency significantly.

During the Spring Festival, technology companies such as Baidu, Alibaba, and SenseTime announced that they will be releasing related algorithms and computing power for virus structure sequencing. They contribute to the fight against the pandemic by empowering medical care with artificial intelligence capabilities.

On 31 January, Baidu Research Institute announced that it will open the linear time algorithm LinearFold and the world's fastest RNA structure prediction website to various genetic testing institutions, pandemic prevention centers, and scientific research centers around the world at no cost. Using this algorithm, the whole genome secondary structure prediction of the new coronavirus can be shortened from 55 minutes to 27 seconds, and the sequencing efficiency will be increased by 120 times, which significantly saves waiting time.

On 1 February, Zhejiang Provincial Center for Disease Control and Prevention launched an automated whole-genome detection and analysis platform. Based on the artificial intelligence algorithm developed by Alibaba Dharma Institute, it can shorten the usual several hours taken for genetic analysis of suspected cases to thirty minutes. At the same time, it can accurately detect virus mutation. SenseTime also coordinated the support of the supercomputing team immediately after receiving the urgent needs relayed by the National Supercomputing Shenzhen Center, providing free high-performance computing resources and supporting researchers in large-scale screening of drugs against the new coronavirus. They concurrently carried out the work on predicting the virus mutation.

In addition, Alibaba Cloud and GHDDI, the global health drug research and development center, cooperated to develop artificial intelligence in drug discovery and development and big data platforms for the conduct of data mining and integration of coronavirus historical drug research and development. At the same time, they announced that during the pandemic, all artificial intelligence computing power will be freely available to public scientific research institutions globally to accelerate the development of new coronavirus drugs and vaccines.

During the drug discovery stage, artificial intelligence mainly plays a role in two scenarios, namely target screening, and compound synthesis and screening. Target screening, one of the cores of new drug development, refers to the discovery of biological pathways and proteins that can slow down or reverse human diseases. In this regard, artificial intelligence uses natural language processing technology (NLP) to undergo deep learning of massive medical literature and process massive amounts of related data, followed by discovering the relationship between compounds and diseases. The target's discovery cycle is shortened as a result of its discovery.

Compound synthesis and screening refers to the combined experiments of millions of small molecule compounds to study compounds with a certain biological activity and chemical structure for further drug development. Artificial intelligence can simulate the drug properties of small molecule compounds so that the best analog compounds can be selected and synthetic experiments can be completed within a few weeks to quickly filter low-quality compounds while enriching effective molecules.

6.3.2 Intelligent Temperature Measurement and Image Recognition

We know that an obvious symptom of the new coronavirus is the increase in a patient's body temperature. Therefore, during the pandemic prevention and control period, testing the population's body temperature has become a key step in assessing infection. However, real-time monitoring of body temperature in public open areas is a huge task. Given the population's rapid mobility and limited manpower and equipment during the pandemic prevention and control period, artificial intelligence monitoring of body temperature has resolved this issue well.

Artificial intelligence body temperature detection mainly uses "human body recognition, portrait recognition and infrared/visible light dual-sensing." Persons with suspected fever are identified through the camera in batches. In this way, high-efficiency identification in public places with large areas and persons with face masks on can be realized. We do not need to inefficiently queue to take our body temperatures, which greatly alleviates contact safety problems and detection efficiency problems during the pandemic.

Medical imaging is an important auxiliary science in the medical industry. However, traditional medical images are interpreted by doctors, resulting in slow diagnosis and a large demand for professionals in related fields. At the same time, the shortage of talents is a major problem, and heavy manual work can lead to misdiagnosis and missed diagnosis. Artificial intelligence landing image recognition can help to resolve such challenges in the medical imaging field.

Artificial intelligence CT equipment equipped with Tencent's artificial intelligence medical imaging products, namely Tencent Miying and Tencent Cloud was deployed in many Hubei hospitals during the pandemic. In general, a chest CT produces an average of 300 images, and it takes 5–15 minutes for a doctor to view the film with the naked eye. With this equipment, the artificial intelligence algorithm can help doctors identify new coronavirus in as little as 2 seconds, greatly improving efficiency while reducing their workload and allowing patients to receive more timely treatment. This effectively alleviated the situation of a severe shortage of medical resources during the early stages of the epidemic.

6.3.3 Intelligent Anti-Pandemic Robot

When the number of confirmed cases is increasing, frontline medical resources are extremely limited. Apart from being understaffed, the overwhelming workload has greatly increased the risk of infection for medical staff. The pandemic must be prevented and controlled, and doctors and nurses also need to be protected. The response plan of intelligent robots to replace medical staff "on duty" can better help medical institutions fight the battle against the pandemic.

During the pandemic, many artificial intelligence companies have successively launched intelligent robots to undertake a large number of simple but labor-intensive process-based tasks such as pre-diagnosis, room inspection, and delivery. These measures can effectively reduce the workload of medical staff, reduce the risk of doctor-patient cross-infections and save medical resources.

Apart from hospitals, the cleaning, and disinfection of public places such as airports and fire stations have also become key landing areas for intelligent robots. The intelligent floor scrubbing robot developed by a local technology company can provide cleaning and disinfection for public places. The "man-machine separation" ensures that workers stay away from crowded places and highly polluted sources.

In addition to offline "postings," the application of intelligent robots extends to online as well. In response to the pandemic, companies such as iFlytek, Baidu, and Yunzhisheng urgently launched an "intelligent voice outbound platform" to assist grassroots communities in carrying out pandemic investigation and related work. From one-to-one phone calls, information collection to report formation, up to 5000 phone numbers can be made in an hour, which is

1000 times more efficient than manual calls. In addition, intelligent epidemic robots developed by some companies can answer online questions about the pandemic, medical attention, and protective measures. The resolution rate of user consultation exceeds 90%.

Due to continuous advancements in computer vision, positioning and navigation, speech recognition, and semantic understanding, the application of artificial intelligence for the use in "liberating repetitive labor" has matured. Self-service consultation, call inspection, and unmanned food delivery. Intelligent robots provided a certain level of manpower and efficiency enhancements and became a solid line of defense in pandemic prevention and control.

6.4 5G Empowers Telecommuting

The pandemic has accelerated the public's acceptance and adaptation to online models.

During the pandemic, we have been practicing the largest telecommuting in history. Traditional business activities, sales, contract signing, customer service, and other key links have also begun to actively explore online versions such as online contracts. Internet medical treatment is also on the rise. The development of remote diagnosis and treatment to remote monitoring has unleashed the potential of the cloud industry and enabled long-term telemedicine.

6.4.1 Readiness of the Office Revolution

Many companies, such as PingCAP, did not have mature localization toolkits during their earliest days of telecommuting. Many of their products were renowned foreign products.

After 2015, the enterprise collaboration platform DingTalk, an online document product Shimo Wendang, and code development platform CODING and other products were launched one after another, bringing about enriching tool-related products. As the pandemic affects the centralized office model, the online office has gradually matured. Telecommuting finally became the most preferred alternative for many companies, while remote office ushered in a new "turning point" of development.

Under special circumstances, online office software, which was once only used as an auxiliary tool, has now become a necessity. Many companies lacking online office experience have consulted various remote office products and services. After WeLink, a subsidiary of HUAWEI CLOUD, announced that it will be free, 5,000 new enterprise registrations were added on the same day.

In addition, according to the needs of the Zhejiang Health and Health Commission, Dingding took only 24 hours from planning to launching the product. Specifically, on the second day, the Dingding team listened to product needs and ideas at 8 a.m. and proposed solutions at 10 a.m. After that, DingTalk was used to assign tasks to staff, who were at home, and began building the product. The first round of acceptance of the product was at 6 p.m. on the second day, the second round of acceptance was at 2 a.m. on the third day, and the third round of acceptance was at 7 a.m. As a result, the Dingding team presented a typical case of telecommuting during the Spring Festival holiday.

Coincidentally, companies with enterprise service experience have opened up their collaboration capabilities at critical moments to help companies continue operating normally even when their employees were at home.

For example, HUAWEI CLOUD provides users with free online video conferences for 1000 accounts and 100-party unlimited meetings. Tencent Conferences offers complimentary access to users for up to 100-person meetings with no time limit, as well as a free upgrade to access 300-person conference collaboration capabilities. Xiaoyuyi Lian Technology provides free 100-party cloud video conferences to government agencies, medical institutions, educational institutions, and enterprises across the country. In addition, Feishu, Yunxuetang, Tencent Docs, HKUST Xunfei, Yunwoke, and Teambition, as well as others, have announced many similar measures.

Remote collaboration office software, which progressed from being a backup to a commonly used office tool for information, social interaction, shopping to office work, has ushered in an inflection point for the industries. The Internet continues playing its role as a tool to help companies develop and improve efficiency.

6.4.2 5G Empowers Telemedicine

During this pandemic, 5G has also enabled remote consultations. Although 5G remote consultations can be completed under the previous 4G network or traditional wired network, it is limited by the data transmission speeds, and a consultation's efficiency may be affected.

Compared with the previous mobile and traditional wired networks, the mobility of the 5G network beats the wired connection of traditional remote consultation. Its high speed allows 4K/8K medical images to be shared in time, and the millisecond delay allows remote ultrasound examinations and the implementation of remote surgery to become possible. As a result, the industry generally believes that 5G is the key to telemedicine development.

A 5G remote consultation uses 5G technology to transmit medical information, followed by remote experts instructing doctors in the diagnosis of a patient's condition. All 5G networks can greatly improve the quality of information transmission and meet the application needs of 4K high-definition audio and video and AR/VR technologies, thereby optimizing consultation objectives. During the pandemic, the country was in isolation. The high-definition and low-latency characteristics of the 5G network were able to efficiently and quickly support the remote sharing of video images and files.

During the prevention and control of the pandemic, medical institutions remotely share data through cloud-to-cloud video conferences, synchronize patient data, realize remote collaborative consultations with multiple experts in multiple locations and actively coordinate high-quality medical resources to diagnose medical conditions. By using 5G remote consultations, local hospitals were able to provide remote consultation services for patients or suspected new coronavirus cases across the country.

After hearing the opinions of the expert team, medical staff can use 5G technology to treat or observe patients in isolation, which plays a key role in the prevention and control of the pandemic. The remote consultation system can connect to the ward, the medical care center, and experts' offices, while experts diagnose and treat patients without entering the isolation area. It can be said that 5G has become a need for the diagnosis and treatment of the new coronavirus. During the pandemic, the medical expert group was led by an Academician.

Zhong Nanshan conducted the first remote consultation on 5 cases of severe and critically ill patients with the new coronavirus in Guangdong through the Guangdong Telemedicine Platform.

In remote consultations, the use of 5G telemedicine trolleys makes remote consultations more efficient during the pandemic. In February 2020, the "5G Telemedicine Trolley" was officially used in Wuhan's Huoshenshan Hospital. Experts beyond Hubei province could use the telemedicine system to diagnose patients in the Wuhan quarantine area remotely, and the treatment effect and speed were further improved.

By installing China Mobile's cloud video client on the medical trolley terminal equipment, the frontline medical staff of Houshenshan Hospital could transfer local medical data (including CT images and detection indicators) via the cloud video and remote desktop function for sharing with Beijing's 301 Hospital to enable remote diagnosis by experts. During telemedicine consultations, medical experts from both sides needed to share patient medical files through auxiliary code streams for them to conduct high-quality telemedicine diagnoses.

The 5G telemedicine trolley has powerful functions, which contributed greatly to the smooth implementation of remote consultation during China's pandemic prevention and control period. It can classify and store a variety of medical devices and drugs, while its convenience provided a more efficient working mode for nurses and improved the efficiency of medical care. In addition, the trolley was equipped with information systems such as HIS, EMR, and LIS in the hospital, which met the needs of doctors for mobile rounds, bedside film readings, and issuing medical orders, while also realizing publicity and education.

This 5G technology-based telemedicine equipment leverages China Mobile's cloud video system to fully realize its high-quality, professional-level video conferencing capabilities. Relying on 5G's large bandwidth characteristics, the 5G telemedicine cart can carry out mobile remote audio and video interactions anytime and anywhere. In addition, mobile ultrasound, defibrillation monitors, and other equipment could also be placed on it to carry out mobile detection, realizing bedside examination and vital sign data collection.

6.5 Wearable Devices during the Pandemic

Wearable devices are devices launched by embedding technologies such as sensors, wireless communication, and multimedia into daily wearables such as glasses, watches, bracelets, clothing, and footwear. Nowadays, smartwatches and bracelets can be seen everywhere, but the development of wearable medical care has not been emphasized and popularized to date. We are not that familiar with them yet.

In reality, in the past few years when the medical device and smart technology industries have been booming. Yet, the wearable medical device industry, which lies between the medical device and smart technology industries, has not been able to fully develop. However, the global pandemic has presented an unprecedented opportunity. Faced with an uncertainty of the world's viruses and epidemics in the future, the pandemic will give rise to a new wave of wearable medical equipment.

6.5.1 Wearing of Self-Cleaning Masks

Given the pandemic situation, mask waste has become a major potential environmental hazard because existing regulatory systems such as China's GB 19083-2010 "Technical Requirements for Medical Protective Masks," YY 0469-2011 "Medical-Surgical Masks," YY/T 0969-2013 "Disposable Medical Masks," as well as the US FDA and EU CE mask standards, are solely focused on their protective performance without any special requirements for their materials.

However, the existing mask materials are polypropylene, meaning they are difficult to degrade. If they cannot be reused, it will exert huge pressure on the environment. It is conceivable that if we calculate the current rate of changing one to two masks per day, the whole world may soon fall into another "white empire" environmental pollution crisis.

To use masks many times, many "witty" netizens have thought of many solutions, such as cleaning, alcohol disinfection, microwave disinfection, ultraviolet light irradiation, steaming in a steamer, and so on. But for the new coronavirus, what we need to care about the most is the mask's filtering for droplets and viruses. The damage to materials caused by the cleaning and disinfection process, such as mask deformation, aging, and loosening of headwear parts, is unpredictable. It may cause a decrease in filtration efficiency and damage to the filter layer, which will eventually cause the mask to fail in meeting air tightness requirements.

If disinfectant agents are used for disinfection, there will also be a problem with residual ingredients. In the current situation, the latest US CDC's recommendations for extended and repeated use of masks show that the most effective way to reuse them is to store them reasonably and heat them under low humidity. This means that steam sterilization is unreasonable as water vapor will reduce the mask's filtering of droplets. The more recommended way is to disinfect with heat from sunlight.

In May 2020, a group of scientists from the Technion-Israel Institute of Technology announced that they have invented a very simple method to reuse masks. It can draw power provided by mobile phone chargers for self-cleaning, and use the built-in carbon fiber layer to kills pathogens such as the new coronavirus.

Team leader Professor Yair Ein-Eli declared that the repeatable method is simple and convenient, and can significantly alleviate problems such as insufficient mask production capacity and environmental pollution during the pandemic. Masks are not entirely non-recyclable. If a simple method can be used to kill the virus on the mask, its lifespan can be significantly extended.

To regenerate a mask, it is necessary to kill the virus on it and ensure that its protective performance is not undermined. The new coronavirus is sensitive to high temperature and ultraviolet rays, and disinfectants containing 75% alcohol content and ether can also effectively kill the virus.

For this reason, many research institutions are trying to extend the life of masks. In February when the pandemic was the most serious in China, Academic Chen Jianfeng of Beijing University of Chemical Technology's team proposed a simple and effective method for regenerating masks. Firstly, sterilize the disposable mask in hot water at 56°C for 30 minutes.

After that, the mask is blow-dried and charged with household appliances such as a hairdryer. As long as the mask can absorb paper scraps, it can be confirmed that it has recharged with the electrostatic charge essential for intercepting the virus and restoring its original function.

The reusable mask developed by the research team of the Technion-Israel Institute of Technology is one of the reusable masks requiring the least equipment and is the simplest method to use.

The secret is that they added a layer of non-conductive carbon fiber material, which can generate heat after being energized, to the N95 mask. They spread this fiber evenly and parallel across its middle layer. When it is connected to the USB cable, the heating fiber will generate a temperature of about 70°C due to the electric current.

While this temperature does not destroy the mask's structure like a steamer, oven, or microwave, it is enough to kill the new coronavirus attached to it. Also, within half an hour, the mask can be disinfected and reused.

Researchers believe that while the number of heating cannot be unlimited, doing so dozens of times should not be a big issue. Researchers from the Israel Institute of Technology have filed a patent application for this device in the United States and hope to launch it in the market at a price of about $1. Taking into account its reusable characteristics, the cost increase is not big. Its greater significance is that it can greatly alleviate the deficiencies of masks while reducing the impact of a large number of disposable masks on the ecological environment.

Smart masks have always been an important trend and opening in the smart wear industry, especially when the environment continues deteriorating and biological genetic engineering continues venturing into new areas. It is difficult for people to predict what kind of super virus will appear in the future. This new coronavirus is just a starting point and not the end of a super virus. Masks will become our daily necessities in the future. The current disposable masks will not only cause a lot of waste, but they will also be ineffective against viruses and health monitoring.

Whether it is the latest research from the Institute of Technology of Israel or the research of other companies, it can be foreseen that smart masks will solve and provide assistance, and become the next major industry.

6.5.2 Clues from Smart Ring Infection Tracking

When SARS broke out in 2003, the Internet was in its infancy. Its infrastructure was not perfect, and big data was yet to gain popularity. Seventeen years have passed. When the pandemic returned, the Internet's environment has undergone tremendous changes, and we have entered the era of big data.

The era of big data favors wearable devices, allowing their data to be updated in real-time, which can be used in the medical field. Undoubtedly, they have great potential for monitoring and detection.

During the pandemic, researchers tried to determine whether a smart ring could help them predict the outbreak of new coronavirus among medical staff. The device was the $299 Oura Ring.

Unlike other wearable devices and smartwatches, this ring can monitor some additional health parameters and capture body signals, such as resting heart rate, (heart rate variability), body temperature, and calorie burn. The most prominent one is the fever sensor, which can detect changes in body temperature. Most of us already know that fever is a common symptom of the new coronavirus.

The research on smart rings provides a new technical perspective for the fight against the pandemic, which is expected to help predict the symptoms of the new coronavirus. Oura Health conducted this research in collaboration with the Rockefeller Institute of Neuroscience at West Virginia University and the West Virginia University School of Medicine. They enabled Oura Ring to not only measure the incidence of increased body temperature in physical symptoms but also to observe the individual as a whole, combining physiological measurements with psychological, cognitive, and behavioral biological measurements, such as stress and anxiety.

In the case of asymptomatic infection, this holistic approach can provide an early and more comprehensive assessment, tracking the physical and mental connections and balance status in the context of asymptomatic infection. Through this analysis, the team can predict the occurrence of fever, cough, fatigue, and other physical symptoms related to viral infections.

Pandemic prevention is a race against time. Since the outbreak, China has taken measures to lock down the city in Wuhan at an alarming rate for the first time. On Lunar New Year's Eve, the construction of the Huoshenshan Hospital was announced, and it took only ten days for it to be declared completed. The construction of Fangcang Hospital began immediately thereafter. More than 10,000 medical professional teams from various provinces are headed to Wuhan to provide support. The military dispatched 20 large transport aircraft to provide transportation for supporting teams and materials. Scientific research and medical research departments raced to isolate virus strains and develop vaccines round the clock.

At the same time, provinces across the country initiated a first-level response to major public health emergencies. During the Spring Festival, citizens isolated themselves at home voluntarily and adhered to pandemic prevention measures by not visiting relatives and going out, wearing masks, and assigning one person to go out or make purchases in the community.

Today, technologies and services, which were part of pandemic prevention efforts, have become very popular. However, there were few policies and measures to use technology to improve the lives of citizens. In the post-pandemic era, the question of how to leverage technology to enhance our lives is something we need to continue thinking about today and in the future.

6.6 A City that Can Resist Infectious Diseases

In the past, living in cities may reduce people's life expectancy. Today, the situation has improved due to technological advances, but busy urban centers remain as key epidemic areas. If there are no quick and effective public health measures to manage the spread of the virus, the larger a city and the closer its connections, the faster the spread of infectious diseases could be in the era of globalization.

In the 21st century, new viruses such as SARS, Middle East Respiratory Syndrome, Ebola, avian flu, and swine flu have appeared. We are currently facing the new coronavirus. If we enter the pandemic era again, what would our city be like, and what kind of city can withstand infectious diseases?

6.6.1 From Space Transformation to Twin Construction

Historically, metropolises have always been regarded as the magnet of mankind's greatest wisdom. However, it is also the gathering place of our oldest enemy – bacteria.

The 1918–1919 influenza was an urban infectious disease pandemic, which was most costly, and its consequences were most dire for mankind. It originated from three places known as major cities at that time, namely Liberty City in Sierra Leone, Brest in France, and Boston in the United States.

Using the United States as an example, influenza spread rapidly from its point of origin to densely populated cities. On 12 September 1918, the first case was reported in DeVance, Massachusetts on 12 September 1918. Within just 6 days, the number of patients surged to 6,674. By the 11th day, which was 23 September, 12,604 soldiers in the city had the flu.

By October, according to just the US military statistics, 20% of the officers and soldiers had the flu. By the end of the flu in 1919, 500,000 people in the United States had been killed by the flu, with urban residents forming the bulk. For example, New York had 60 deaths per 100,000 people, while Philadelphia had 158 deaths per 100,000 people.

The new coronavirus first broke out in Wuhan, China, which was the most densely populated city in central China with a huge population of over 11 million. In the United States, where the pandemic is severe, New York, its most densely populated city, was also badly hit by the pandemic.

Transforming urban space has become an urban planning problem that we must solve in the future.

The traditional urban spatial form often starts from the overall urban planning layout, which generates corresponding land use and triggers the flow of people in the city. The advancement of information technology has accelerated the time and space exchanges of knowledge, technology, talents, and capital, which resulted in the continued expansion of urban production and increased complexity in the types of activities. This has promoted industrial restructuring and spatial reorganization to some extent, which in turn changed the spatial pattern of regions and cities. "Flow Space" has become the main carrier of regional, city, and residents' activities.

The rapid development of the Internet and big data has promoted innovation in the fields of economy, finance, and social services, and provided support for the multidisciplinary integration of urban research. In addition, the use of big data can shed light on residents' travel behavior, traffic operations, and spatial development, thereby promoting the dynamic management of urban planning. With the emergence of big data, urban planning assessment has shifted from a physical space assessment with land use as the core to a comprehensive urban socio-economic assessment with individual daily behaviors at its core.

Urban space arises with planning. To build a city that can resist pandemics, it is necessary to transform the urban space. The use of big data and new technology to dynamically identify urban spaces makes it possible to provide more activities for people even in a densely populated area.

At the same time, the popularity of digital twins has continued rising in the past few years, frequently appearing in the speech topics of major summit forums, and has attracted attention from those who are within and outside the industry. Simply put, a digital twin is to create a digital version of a "clone" with a device or system serving as a basis. This "digital clone" is created on the information platform and is virtual. The biggest feature of the digital twin is that it is a dynamic simulation of physical objects. In other words, the digital twin is "mobile."

The outbreak was a serious test to China's push to modernize its national governance system and governance capabilities, especially the governance problems and contradictions in megacities such as Wuhan, which were fully exposed. How to build a complete mega-city governance system and a modern smart governance capability system is a test for these mega-cities.

In addition to "mobility," digital twins feature the characteristics of a "full life cycle." In the field of industrial manufacturing, there is a term called "product lifecycle management (PLM)." The entire life cycle of a digital twin can run through the entire cycle of product design, development, manufacturing, service, maintenance, and even scrap recycling. It is not limited to the creation of better products. It also includes better use of products in later stages.

From the perspective of "full-cycle management," a city is an aggregation system that integrates multiple elements and problems, and urban governance is a systematic project. In the face of the pandemic, the previous urban governance methods involving taking stop-gap measures seem to be losing their effectiveness.

With the massive concentration of population and the acceleration of urbanization, the security risks of cities are becoming increasingly complicated. Because urban disasters have the characteristics of multiple occurrences, diversity, complexity, superposition, derivation, and conductivity, the governance of urban disasters must be managed systematically and at the source(s).

The urban smart governance system in the new era should form a closed management loop of warning and decision-making in the beginning stage, followed by response and execution in the middle stage, and ending with summary and learning in the final stage. In such a closed loop, the beginning, middle and final links are closely intertwined. All urban departments must have clear rights and responsibilities and scientific cooperation to ensure the smooth flow of information.

Therefore, in the process of building a smart urban governance system, be it to solve urban operation problems or resolve risks, it is inseparable from a more efficient top-level design of urban modern governance. We need to start from the top-level design of urban construction, build digital twin cities in the new era based on digital twin technology, realize the advanced development model of intelligent urban governance and sustainable operation, and advance digital urban governance in detail.

6.6.2 From the Improvement of Urban Sanitation Systems to Urban Automation

With the construction of "digital twin cities" at the forefront, we will further promote the extensive application of urban spatiotemporal big data, cloud computing, and artificial intelligence in urban governance, and vigorously develop "Internet + governance" and "intelligence + governance." The usage of data twins will empower urban smart governance and realize the closed-loop management of urban smart governance, as well as comprehensively improve the level of "full-cycle management" in urban governance.

Not all cities are vulnerable to diseases. Rich cities like Copenhagen have plenty of green spaces and cycling facilities. But for those living in economically underdeveloped cities like Nairobi in Kenya, or Dhaka in Bangladesh, the situation is completely different.

Elvis Garcia, a public health expert, and lecturer at Harvard University, said that places without proper sanitation facilities and clean water are where infectious diseases are most likely to begin and spread.

In the 19th century, cholera was said to be a global nightmare at the time. About 130,000 people died of cholera in Britain alone. What is even more frightening is that more than 20 million people died of cholera in India.

The outbreak and spread of cholera are not only related to medical technology but also the result of comprehensive social governance. Water pollution and deterioration of the sanitation environment have directly caused disasters. In the process of fighting cholera, Western European countries have gradually established effective infectious disease prevention and control mechanisms, and relatively complete public health systems.

The progress of modern civilization and the development of science and technology have qualitatively improved the urban sanitation environment and sanitation management. However, the former continues to reveal many contradictions in the face of the pandemic.

Public health emergency management capabilities test the governance capabilities of modern international metropolises. The concept of public health includes far more than just medical and health care. It requires the integration of cross-departmental, cross-level, and cross-regional information to jointly achieve optimal results.

However, there are still many problems in information sharing to date. Each hospital is an "island," and even a lot of software in a hospital cannot be shared. "Information islands" are not a technical problem, but institutional problems, a reflection of the fragmentation of the medical service system.

It is foreseeable that future urban construction will develop towards a more ideal health system. A framework for performance evaluation of the health system of megacities will be established. Internet hospitals spawned by the pandemic will largely share the pressure on medical resources of physical hospitals and the development of big data. The landing of blockchain will also break the barriers of medical information.

On the other end, automation has become the trend of social development and has gradually entered people's daily lives. Although there is still a long way to go before the popularization of a sound health system, the outbreak of the pandemic is undoubtedly a good opportunity to realize automation.

Restaurants today are already working on automating part of the delivery process. Given the pandemic situation, more companies will push the automation limits in their services. To survive, some companies must automate, especially those which intend to try automating their chain and payment services. When automation is successfully implemented in the company, there will likely be no need to hire people to hold these positions after the pandemic is under control.

Challenging moments in history present opportunities for innovation. During the pandemic, data analysis, artificial intelligence, and robotics play an important role. Data analysis and artificial intelligence are used to develop drugs to help assess the spread of the pandemic, while robots can help medical teams to carry out their work remotely and safely, which was previously unattainable.

There are countless examples of these technologies changing the way the world responds to the pandemic. Since the outbreak started, demand for robots that test patients with ultraviolet radiation has increased. Doctors have used artificial intelligence to check infected patients, while companies have realized the automatic prediction pertaining to the spread of new coronavirus.

With the promotion of 5G, more 5G applications are being utilized. The 5G network provides a higher transmission rate, precise low-latency control, and positioning, which will help the implementation of unmanned driving, realizing the information fusion of vehicle and roadside perception. This reduces the computational complexity of the vehicle-mounted system and effectively resolves vehicle-vehicle or vehicle-road coordination issues.

The urban roads of the future will be intelligent digital roads, which are coded at every square meter. Using active radio frequency identification technology (RFID) and passive radio frequency identification technology (RFID) to transmit signals, intelligent traffic control centers, and cars can read the information contained in these signals. Moreover, RFID can accurately locate underground roads and parking lots.

Undoubtedly, the pandemic is another big step forward towards city automation.

7

Crisis Gives Rise to Opportunities, with an Imminent Turning Point

7.1 The Wind of "De-sinicization" is Everywhere

The global shutdown brought about by the pandemic has affected every aspect of social life. According to market predictions, the impact of the pandemic on the global economy will surpass the 2008 Financial Crisis, even rivaling the Great Depression of 1929. As the pandemic intensifies, there is a lot of discussion about how it will affect China's socioeconomic politics. Among these, the notion of "deglobalization" is particularly striking.

The issue of "deglobalization," hotly discussed in Chinese public opinion circles, is closely related to the decoupling between the West and China, or rather the "de-sinicization" of Western countries. The concern about "de-globalization" is related to the negative effects of "de-sinicization" on China's interests, and this concern is very much justified. However, in today's international economy based on the Internet, globalization has become an unstoppable trend in the course of human development. No country can survive on its own by leaving the globalized economic network. Due to the integrity of its industrial chain and manufacturing advantages, China plays an indispensable and irreplaceable role in the global supply of goods.

7.1.1 Post-pandemic Issues of Micro, Small and Medium-sized Enterprises

At present, in addition to some large enterprises, especially the badly hit aviation industry and the energy industry, small, medium, and micro-enterprises around the world have also been badly impacted by the pandemic. In addition to macroeconomic and monetary policies, governments around the world are helping SMEs to cope and survive the crisis while transforming these businesses. This shall become a major issue facing SMEs all over the world. Although the pandemic has caused a significant blow on economic growth and the labor market in the short term, it has also provided new opportunities for the strong growth of new economy industries, as represented by the digital economy.

The special circumstances of the pandemic have enhanced the adaption between customer demand and model innovation, ushering a new leap in the digital economy. From competing for website monetization to uncovering new demands, this pandemic will have a long-term impact on businesses. SMEs experiencing cashflow pressures during the pandemic have begun to actively migrate their businesses online to reduce uncertainty in business growth.

Meantime, shared manufacturing will become a new production method that aids businesses to cope with uncertainty. Shared manufacturing is a new production method that gathers scattered and idle capacities during each stage of the production while it carries out elastic matching and dynamic sharing among demanders. In essence, it uses digital technology to enhance the flexibility of economic activities.

On the one hand, shared manufacturing platforms have become a key link in promoting the development of the shared manufacturing model, gradually breaking the boundaries between businesses, and promoting the development of production organization to a networked organization and a platform-based economy. On the other hand, shared manufacturing not only promotes the evolution of technology and capital input but also changes the labor input of businesses, making shared labor an inevitable trend in company employment.

In the short term, the pandemic has caused an economic downturn, severely impacted social life, and threatened the survival of companies. From a longer historical perspective, the pandemic resulted in a unique depression scenario where both supply and demand are weak, but it has also accelerated the difficult process of business transformation. Understanding how small, medium, and micro enterprises seize new opportunities in the digital economy is the key to solving problems after the pandemic.

There is no doubt that micro, small and medium-sized enterprises play an important role in the economic development of various countries. These businesses not only contribute a large amount of tax revenue but also guarantee employment among the vast majority of the population. They also contribute to economic development and social stability. Therefore, the survival of small, medium, and micro enterprises during the pandemic is one of the biggest challenges that countries must strive to solve. Dealing with the crisis and achieving transformation are major issues faced by small, medium, and micro enterprises of all countries.

Beneath a crisis lies another crisis. However, each crisis will accelerate the formation of trends and generate new business opportunities. The SARS epidemic increased the popularization of the Internet in the past, while COVID-19 shaped the direction of humanity in the throes of pain. This pandemic is not just a source of stress, it also brings opportunities to individuals and enterprises, and even to entire societies and countries.

7.1.2 The Vulnerability of Supply Chains Under Globalization

Compared with the early days of industrialization, today's increasingly complex and networked global supply chain may make them less resistant to risks. The highly globalized industry chain appears fragile in the face of the pandemic.

Take the global automobile industry as an example. Two weeks after Wuhan's citywide lockdown, Hyundai's motor plant in South Korea suspended production. The reason is that a parts supplier in Qingdao had less than 15% of its workforce resuming work after the Spring Festival, forcing the South Korean car factory across the water to stop production.

Beneath the fragility and complexity of the supply chain are productive and economic benefits. Economies have different comparative advantages in production, sales, and other aspects. In the case of the iPhone, its screens and cameras are made by suppliers in South Korea and Japan, its phone components are assembled in China, and its stocks are listed in the United States. Apple products are sold all over the world, benefitting every country in the industry chain and the capital funding it.

China's status as the workshop of the world is attributed to its lower overall costs. Over the past two decades, China has become a major player in the global supply chain as Western companies entered China on a large scale to reduce production costs and open up new markets. Today, China's GDP accounts for nearly 20% of the global GDP. According to a survey conducted by a German company, four out of five big companies are reliant on Chinese suppliers. Arguably, China's status as the world's factory is not just anyone's wise plans or policies of visionary leaders, but the result of organic growth over decades.

China has not just been a major producer of face masks for a long time, but it has shown explosive growth in its production capacity during the pandemic. From early February to early March, the daily production of masks climbed from 10 million to 100 million in just a month. In the pharmaceutical field, China accounts for 60% of the global production of APIs, influencing the production of the world's largest pharmaceutical companies. China's capacity for producing ventilators is relatively small, but it still accounts for one-fifth of the world's production.

The pain caused by the epidemic coupled with the long-term lack of political mutual trust has caused anxiety in Western countries. They are confronted with the fragility and complexity of supply chains, and they want to move part of their production capacity out of China in the future to diversify their risk. However, the reality is that China is the first country in the world to fully resume production and it has the advantage of having a relatively complete supply chain. These factors, to some extent, shall continue to promote the globalized industrial model.

On March 31, 2020, Trump stated at the Coronavirus Information Conference in the White House that he would "turn the United States into a fully independent and prosperous nation with full sovereignty over energy, manufacturing, economy and national borders." Then Larry Kudlow, director of the White House National Economic Council, suggested that American companies could be encouraged to move back to their home country through repatriation spending.

Before March, in response to the pandemic, Japan's Ministry of Economy, Trade, and Industry has earmarked 243.5 billion yen (about RMB 15.8 billion) from the "Supply Chain Reform" program. This amount shall fund Japanese manufacturers to withdraw their production lines from other countries to diversify their production bases and avoid relying too much on overseas supply chains.

However, be it the United States or Japan, their industrial and economic models have shaped today's industrial and economic structure in the world through globalization for nearly half a century. China has also defined its unique position in the global industrial division of labor for nearly fifty years while making important contributions to the world economy. It is difficult for any country to overturn the economic and industrial model built by the joint efforts of the world's major economies over the past five decades, let alone through a single ideal or slogan from politicians.

7.1.3 The Effects of the Pandemic on Traditional Businesses and Cross-border E-commerce

The spread of the pandemic has further weakened the sluggish global trade. Coupled with severe restrictions on the cross-border movement of people and the cancellation of traditional face-to-face business transactions, it is difficult for businesses to rely on offline exhibitions to expand to international markets. Since the COVID-19 outbreak, more than 100 major international exhibitions have been postponed or canceled, such as the World Mobile Expo in Barcelona and Cologne Hardware Fair in Germany, impacting the ability of export-oriented companies to expand to international markets. However, it is precisely because the B2B culture that emphasizes face-to-face transactions has become the straw that broke the camel's back, there is a further interest of export-oriented companies to grow their businesses online.

On March 6, 2020, the Zhejiang Department of Commerce and Vietnam's Ministry of Trade and Industry jointly held the launching ceremony of the Zhejiang Export Commodities Online Fair (Vietnam station). From 15th June to 24th June, the 127th China Import and Export Fair was held online for the first time, making this spring canton fair an unprecedented event.

From June 8th to June 28th, China's technology giant Alibaba's Ali International Station also held its online exhibition, named the Internet Fair. This cross-border B2B e-commerce platform eventually attracted more than 10 million buyers from all over the world. Alibaba revealed that the Internet Fair led to a year-on-year increase of 177% in the number of inquiries, a year-on-year increase of 243% in paid orders, and a 124% increase in actual transactions.

This online exhibition turned the long-cherished vision of Alibaba founder Jack Ma to "buy and sell the world" into a reality. Figures released by the General Administration of Customs on July 14th showed that in the first half of 2020, while China's overall exports fell by 3% year-on-year, exports through the customs' cross-border e-commerce monitoring platforms rose by 28.7%.

China's Ministry of Commerce stated in January that private enterprises accounted for 43.3% of imports and exports in 2019, making them the country's largest importer and exporter for the first time. Among them, exports by private enterprises accounted for 51.9% of the total exports, indicating that half of the exports in the country are undertaken by private enterprises.

In addition, China's economy contracted by 6.8% year-on-year in the first two quarters of 2020 and grew by 3.2% year-on-year in the second quarter, demonstrating how China is emerging from the shadow of the pandemic. During this period, imports and exports of private enterprises also achieved growth despite the overall trend, thus playing a more prominent role in the stabilizing of foreign trade. China's General Administration of Customs reported that private enterprises show positive growth of 4.9% in the first half of the year, even as the total value of China's imports and exports fell by 3.2% year-on-year.

This also allows people to see the benefits of the digital economy on private enterprises that engage in cross-border trade in the post-pandemic era. With the help of digital methods, companies can find overseas buyers more conveniently and intelligently, while understanding the fundamental needs of buyers more accurately. In turn, products are repeated and updated accordingly. Digitalization can also significantly reduce the cost of transaction compliance.

Besides traditional trading companies, cross-border e-commerce also ushered in growth opportunities. In traditional international trade, the flow of goods, technology, capital, and other elements are mainly concentrated among large companies. With the development of digital trading platforms, small and medium-sized businesses are empowered to engage in vertical segmentation through cross-border e-commerce trading platforms. By providing accurate products and services to satisfy users, digital trading platforms are becoming an important force in promoting international trade.

As of 2020, China's small, medium and micro cross-border e-commerce business has reached more than 200 countries or regions in the world, and its export targets include developed countries in Europe and the United States, Middle East, and Arab countries, Southeast Asia, and developing countries in Africa. With the needs of countries at different stages of development varying greatly, as compared to traditional supply chain systems that are unable to obtain timely changes in market demand, the flexibility of SMEs and the technical support of digital trade have given them a comparative advantage.

As a result of the pandemic, on the one hand, demand for online shopping has increased significantly. Due to the restrictions on the cross-border movement of people, there is a significant decrease in overseas shopping and consumers tend to use online methods to purchase high-quality overseas products. According to data released by Ningbo Free Trade Zone, from January to February, the total number of cross-border e-commerce import declarations issued by the zone exceeded 11.8 million, an increase of 27.8% year-on-year. Cross-border import amounted to 2.4 billion *yuan*, an increase of 42.5% year-on-year. Arguably, the "stay at home" online cross-border shopping has promoted the development of cross-border e-commerce.

On the other hand, the effects of cross-border e-commerce are evident. Compared with traditional export-oriented enterprises, cross-border e-commerce enterprises mainly use online methods to acquire customers, and they are more adept at picking up and mastering digital marketing. With the gradual recovery of international aviation and port logistics, cross-border e-commerce enterprises, through relying on digitalization, drive the flow of international cargo, logistics, and information more rapidly, thus ushering opportunities for rapid development.

7.1.4 Can Supply Chains be Successfully "De-Sinicized?"

The impact of the pandemic on the global supply chain has made countries pay more attention to local suppliers and forced companies to restructure their original industrial chains. At the same time, the shortage of vital goods, such as medical protective equipment and ventilators, has made the developed world more aware of the possible consequences of the "hollowing out" of their manufacturing sector.

In the short term, however, many factors prevent the supply chain from being "de-sinicized."

First, the pandemic has made fighting COVID-19 a top priority for countries around the world, making it impossible to initiate "de-sinicization" at this point. As of March, 90% of masks in the United States are imported from China, and China accounts for about 50% of the world's mask production. Also, there are 21 ventilator manufacturers in China accounting for about 20% of the global production capacity. On the other hand, there is a 50% demand gap for ventilators, masks, protective clothing, and other essential supplies in Europe and the United States.

Essentially, European countries and the United States have to rely on imports from China during the pandemic. At this point, if "de-sinicization" were forcefully implemented, it would no doubt be met with resistance. So, at least until the outbreak slows down and is brought under control in the US and Europe, markets do not need to worry too much about "de-sinicization."

From a long-term perspective, it is very difficult for Western countries to restore their manufacturing industry and form an independent industrial chain system.

First, the US strategy of "manufacturing repatriation" has been in place for a decade, with little results. No matter how hard the US government promotes it, companies are reluctant to invest too much at home if it does not align with their interests.

In the end, the manufacturing sector has been losing ground in the US economy despite repatriation efforts in the past decade. From 2008 to 2015, the share of manufacturing value-added in the US GDP has never returned to the pre-financial crisis level of 12.8%. The level in 2015 was much lower than the level of around 15% before 2008. As a result, the population employed in manufacturing in the United States is also shrinking, and it has been on a downward trend since the beginning of this century. In the three years from 2015 to 2017, the percentage of the population employed in the manufacturing sector in the United States was 8.8%, 8.3%, and 8.5% respectively.

More importantly, the reason why China has emerged as the "world's factory" is because it attempts to adapt actively or passively, in other words, it has undergone different history stages over the last 30 or 40 years. This is hard for other countries or regions to replicate.

At the same time, China has a huge consumer market. It is rational for companies to place their manufacturing facilities near their markets. The profit-seeking nature of capital incentivizes companies to consider their production layout from the perspective of their economic benefits. Therefore, governments can neither force all overseas companies to convert their production layout that does not align with their interests nor can they provide sufficient compensation to help the companies in making the conversion.

Profit is the fundamental motivator for business development. Entrepreneurs will still consider how to maximize the benefits of the industry from their interests. For governments and businesses in Western countries, bringing production back home may bring some new job opportunities in the short term. However, in the long term, it will lead to lower profit margins and higher product prices. This is uneconomical as it will ultimately hurt consumer welfare.

Additionally, the combination of liberal policies and an open business environment, as well as the continuous improvement and upgrading of infrastructure, have made "de-sinicization" difficult to implement.

In any case, COVID-19 has already made a significant impact on a global scale. The global industry and supply chain have been hit hard by the pandemic, and they are bound to face adjustments in the future. Society also faces the challenge of consolidating, complementing, and innovating the industrial chain.

To consolidate the industrial chain, it is necessary to strengthen investment in crucial global industrial chains and enhance China's position in the global industrial chain. Crucial global industrial chains have the following characteristics: First, the international division of labor and integration has to be the most closely knitted. Second, the development of science and technology and the expansion of the market in this field need to be the fastest. Of course, this is where the problem is the most concentrated. Third, competition in this field must be the most intense internationally.

As such, China has to maintain high-quality economic development to dominate high-tech industries and win the fight against the pandemic and hegemonism. It is, therefore, necessary for China to be at the forefront of futuristic high-tech industry technology, possessing independent intellectual property rights, and not be at the mercy of foreign countries. If China wishes to achieve leapfrog development in science and technology, it cannot simply follow and imitate others, let alone wait for technology transfer. Instead, it must strengthen initiatives for independent innovation. Without independent innovation, China lacks the leverage in international cooperation.

In addition, under the broad framework of the Belt and Road Initiative, enterprises should accelerate the process of going global and they must share the manufacturing advantages and industrial chain models accumulated by China over the years with more countries. In particular, through establishing industrial parks and economic corridors abroad, investing in manufacturing projects, and innovating industrial chains, these industrial chains may become an important force and model of the global chain in the post-pandemic world.

Finally, from China's perspective, more efforts should be made for China to open itself up to the outside world and integrate itself into the new layout of industrial chains established by large international companies. At this point, there is a need to perfect the social credit system. After this outbreak, large international companies have to rearrange their industrial chains, which requires them to integrate themselves into the new layout with a very open mindset.

Whether it is from economic, political, or even cultural perspectives, "de-sinicization," broadly speaking, is impossible to achieve. It is a political slogan that goes against the trend of global economic development. The Chinese influence has been spreading over the world for many years. For a long time, American policymakers have been thinking about changing China. However, they have never considered how to work with China to complement each other's strengths while bringing greater benefits to the global economy and the entire mankind.

Today's Sino-US relations, however, seem to have entered a stage unrelated to the economy. If one looks at the global economy rationally, "de-sinicization" is an emotional concept that assumes heavy costs for both countries. This confrontation will persist after the pandemic, which is both a crisis and an opportunity for China.

7.2 Resolving the Predicament of Small, Medium and Micro-Enterprises after the Pandemic

COVID-19 has a huge impact on the global economy and the International Monetary Fund expects the global economy to contract by 3% in 2020. Due to the economic tsunami caused by the pandemic, China's economic growth in the first quarter fell by 6.8% year-on-year, making the survival and development of small and medium-sized enterprises particularly difficult.

The importance of small and medium-sized enterprises in China's economic development is beyond any doubt. In August 2018, during the first meeting of the Leading Group for the Promotion of the Development of Small and Medium-sized Enterprises of the State Council of the People's Republic of China, the famous "56789" assertion about SMEs was made in the meeting, that is, SMEs provide 50% of the country's tax revenue, they also generated more than 60% of the GDP, filed more than 70% of invention patents, provided 80% employment in cities and towns, and absorbed 90% of new employees. It can be said that micro, small and medium-sized enterprises have been the ballast of China's economy since its reform and opening up.

However, due to the impact of COVID-19, in the first quarter of 2020, small, medium, and micro-enterprises in mainland China were affected by shrinking demand and production suspension, causing their operating income to precipitously fall. As a result, many small and micro enterprises were faced with a severe existential crisis.

7.2.1 From Cashflow Woes to Financial Bailouts

The problem of financing SMEs has always existed, but COVID-19 has made this issue even more pressing. According to sample operating data, the operating income of small, medium, and micro-enterprises in the first quarter of 2020 is less than 50% of the same period last year, with more than 80% of SMEs facing cash flow problems.

Among them, SMEs in the education, catering, accommodation, entertainment, culture, and manufacturing sectors are the most affected by the epidemic. The turnover of SMEs in the education sector was only 10.2% and 11.8% in February and March 2020 as compared to the same period in 2019. The turnover of SMEs in the accommodation and catering industry was only 12.8% and 23.5% as compared to the same period last year. The turnover of SMEs in culture, sports, and entertainment sectors in February and March 2020 was less than 30% as compared to the same period last year. The turnover of SMEs in the manufacturing sector in February and March of 2020 was less than 40% as compared to the same period last year.

The decrease in operating income of education, catering, accommodation, sports, and entertainment industries was mainly affected by demand-side factors, including the sharp decrease in demand caused by the pandemic prevention measures and consumers maintaining social distancing. The decline in operating information of the manufacturing industry was mainly due to supply-side factors, including workers unable to return to work and the inability of upstream and downstream (domestic and foreign) supply chains to coordinate work resumption. More importantly, due to reasons such as high operating costs, high labor intensity, single industrial chain channels, and low degree of operation digitalization, SMEs not only were slower in resuming production compared to large enterprises, but they also tended to take more passive measures when dealing with the pandemic.

The direct consequence of the sharp decrease in operating income is that most small, medium, and micro enterprises are facing cash flow difficulties. According to a survey conducted by OneConnect in February, more than 80% of SMEs' working capital was not going to last beyond June 2020.

The pandemic has become the most important variable in the global economic trajectory, and its continuous spread has repeatedly altered expectations among people from all walks of life. The pandemic has impacted almost every aspect of business, be it employment, cash flow, production, or sales.

Due to the lack of sufficient scale and collaterals, small and medium-sized enterprises are extremely short of funds. The prolonged pandemic has caused a significant halt in the offline economy and a sharp decrease in business activities. Small and medium-sized enterprises with tight cash flow are more negatively affected by the pandemic and more prone to collapse.

Helping small and medium-sized enterprises to overcome difficulties and ensuring their survival are effective measures to prevent the rise of social unemployment and the decline of residents' medium and long-term consumption due to income stoppage. In other words, these measures help to prevent the crisis from being transmitted from the supply side to the demand side.

Although the central government has issued a series of policies to help SMEs cope with the crisis, among which a total of more than 6.03 trillion *yuan* have been allocated for the financial support policy as well as the tax and fee reduction policy, small and medium-sized banks still face challenges in issuing loans to small and micro-enterprises. As of March 30, 500 billion *yuan* out of the 800 billion *yuan* bailout fund previously distributed by the central bank have been utilized. Among which, 336.2 billion *yuan* was converted into loans to micro, small and medium-sized enterprises.

Credit funds of banking systems are the main external source of financing for Chinese small and medium-sized enterprises. A large number of studies have shown that large banks tend to provide financial support to large-scale enterprises, while small and medium-sized banks are suited for serving small and medium-sized enterprises. Ayaratne and Wolken's research shows that even in countries with a high degree of financial liberalization like the United States, small and medium-sized enterprises tend to rely more on small and medium-sized banks to obtain loans.

Whether the financial support policy from the central government can fully play its role depends on the lending capacity of small and medium-sized banks. However, in reality, the rise of non-performing loans and the narrowing of the net interest margin still hinder small and medium-sized banks from granting loans to small and micro-enterprises.

In terms of the non-performing loan ratio, the combination of the macroeconomic downturn pressure and the impact of the pandemic has increased the operating difficulties of companies, especially among small and micro enterprises, and made the risk management of bank credit more difficult. Data for the first quarter of 2020 showed that the non-performing loan ratio of the banking sector was 2.04%, with an 0.06% increase from the beginning of the year. Due to the lag in identifying non-performing loans, the real test of bank asset quality may not come until the second quarter.

The strict assessment of the non-performing loan ratio is one of the reasons for preventing SMEs from getting loans. To meet the tolerance requirement for non-performing loans, small and medium-sized banks hesitate to extend loans to micro, small and medium-sized enterprises, especially those in the manufacturing, accommodation, and catering sectors that have been severely affected by the epidemic. Credit officers may also choose to cut back on loans to small, medium, and micro enterprises for fear of being held accountable. Therefore, it is worth thinking about how to implement short-term measures to improve the tolerance of non-performing loans and to flexibly adjust the assessment methods of credit officers. At the same time, the rise in non-performing loans will also put pressure on banks to increase reserves and replenish capital. Capital constraints will also restrict banks from issuing small, medium, and micro loans.

In terms of net interest margins, it will shirk, if small and medium-sized banks issue low-interest loans to micro, small and medium-sized enterprises. At present, the net interest margin of banks using quasi-bank reloans to SMEs is 2.05%, which is lower than the average net interest margin of commercial banks.

Based on this, helping small and medium-sized banks manage the credit risks of loans to SMEs are the key to solving the financing problems of SMEs. The loan risk of SMEs is not only reflected in risk identification and adverse selection before the issuance of loans but also reflected in the supervision of the post-loan funds and the prevention of unethical behavior.

Big data, blockchain, the Internet of Things, electronic invoice technology, and digital identity authentication technology can be used to prevent SMEs from using loan funds for policy arbitrage, transferring funds to the real estate market or capital market, and increasing credit risks of loans to SMEs. The main purpose of using such tech is to give SMEs an accurate credit profile, build low-cost and low-risk real-time credit information service systems for pre-loan evaluation, in-loan monitoring, and post-loan real-time tracking to create a new model of large-scale, fast, and low-risk financing for small and medium-sized enterprises.

For example, through electronic invoice technology, banks can quickly query and verify information in the electronic invoices in batches, thereby speeding up the capital turnover of enterprises. Through digital identity authentication technology and Internet of Things technology, it can monitor abnormal behavior of borrowers in real-time and provide a real-time warning. By building a real-time credit information system, it can not only replace traditional credit reviews while improving credit efficiency but can also develop into in-depth, professional modeling applications. Innovations like invoice loans and credit loans will also become an inevitable trend in the post-pandemic world.

7.3 The Digitalization of Foreign Trade during the Pandemic

From the Silk Road in the 4th century BC with horse-drawn carriages and boats as the main means of transportation to China's entry into the World Trade Organization (WTO) in the 20th century, China's international trade has experienced a long period of development, while its foreign trade business model has also been undergoing constant transformation and upgrading.

From a global perspective, decentralized on-site transactions, represented by the natural economy, can be regarded as the changes in community structure and main activity forms throughout different historical periods. The ancient times, which were characterized by decentralized on-site management, were called the first civilization period. The two industrial revolutions, represented by the market economy and the high degree of centralization, were called the second civilization period. The intelligent co-sharing society, characterized by efficient decentralization, is called the third civilization period, or rather off-site economic civilization.

Off-site economy, the result of smart labor, refers to the degree of intelligence within the time node in the intelligent economic era. It reflects the social and economic changes brought about by the intensification of off-site activities in today's society, as well as a series of economic phenomena and activities that follow thereafter.

This is even though the COVID-19 pandemic has led to the largest human quarantine operation to date. The tide of anti-globalization has brought great uncertainty to China's foreign trade industry, but it has also further promoted the development of an off-site economy, giving China's foreign trade reasons for optimism amid the global economic downturn.

7.3.1 "Buying and Selling the World"

The digital economy has brought in a "new normal" for international trade.

First, with the spread of the Internet, more SMEs can procure goods all over the world, resulting in significant structural changes in global trade. There used to be a few large buyers in the past. However, the increase in the number of buyers caused purchase orders to become increasingly fragmented. This means that there is a need for businesses to engage in stock increment.

An important method for a stock increment is to infer products in demand based on the data from buyers. Although as compared with B2C, export-oriented companies targeting B2B are slower in updating themselves, it is crucial for B2B products to constantly update themselves to maintain sufficient competitiveness, or risk falling in rankings on the platform.

Second, new forms of technology emerge as export-oriented companies embrace digitalization, mainly in the form of digital marketing, virtual marketing, and live broadcast marketing. In May and June 2020, Alibaba International Station held 511 online exhibitions and Internet fairs that featured Class B live broadcasting and other technologies while providing services for international trade supply chains such as translation, customs clearance, tax refund, logistics, and financing.

E-commerce live broadcasting, by relying on referrals, social media celebrities, sale promotions, and other interactive features have grown rapidly during the pandemic.

According to the 2019–2020 Research Report on China's Online Broadcast Industry released by iiMedia Consulting, more than 40% of the interviewed users said they had watched "live broadcast + e-commerce" programs. Alibaba hosted thousands of live broadcasts in their 511 online events. During the three weeks of the Internet Fair, it originally planned to hold 6000 Class B live broadcasts, but it was so popular that more than 8000 live broadcasts were held.

Finally, the notion of "buying and selling to the world" becomes more prominent. From the perspective of sellers, the structural changes of international trade in recent years can be seen in the commodities themselves, with obvious differentiation of goods for people from different countries or groups. Consumers from Europe or America mainly purchase 3C electronic products and fashion clothing, buyers from Southeast Asian countries mainly purchase 3C electronic products, and Russian consumers rely on the export of China's consumer goods. Additionally, 70% of the imports of consumer goods come from AliExpress, and the sales volume of subdivided categories is relatively balanced. In recent years, the export of electronic cigarettes and self-balancing scooters has increased, while the quality assurance of goods has gained increasing attention from foreign importers.

From a buyer's point of view, there are many excellent overseas suppliers but buyers need to clearly define the demand for their goods. In North America, the demand for goods includes vitamins, health care products, special agricultural products, while countries like the United States, Vietnam, and Italy have their unique goods to showcase. Therefore, the development of global trade will move towards the direction of "buying the world."

In the second half of 2020, as the export of pandemic prevention supplies slows down, China's export will mainly rely on the recovery of external demand. Li Kuiwen, Director of the Department of Statistics and Analysis of the General Administration of Customs, said in a press conference, "With the increase in uncertainty and instability in China's foreign trade and the trade frictions between China and the US, the import and export situation remains grim and complex in the second half of the year."

Nevertheless, there is still huge room for China's foreign trade to grow. The data shows that while the digitalization of China's domestic retail market has exceeded 30% at present, the digitalization of foreign trade is less than 10%. Despite the uncertainties of the global economic downturn, the transformation to digitalization can be realized as long as the principles of developing a digital economy are well understood. This transformation is also aided by the West's inadequate pandemic response that severely impacted their production and manufacturing abilities. Therefore, there are still reasons for optimism in China's foreign trade.

7.4 The Pandemic Hits Global Aviation

COVID-19 has caused the worldwide aviation industry to suffer from its most severe blow in history. According to aviation data provider Flycom, as of March 2, 20202, domestic and international airlines have processed refunds for 24.545 million tickets with a total face value of 27.1 billion *yuan* due to the mass cancellation of domestic and international flights. For a sector that relies heavily on capital, the mass cancellation has disrupted cash flow and caused a severe blow to airline companies.

The International Air Transport Association estimates that the global aviation industry needs at least US$150 billion to US$200 billion in financial assistance to overcome the crisis. The recent financial forecast report for the global air transport industry from the International Air Transport Association shows that global airlines will suffer a loss of US$84.3 billion in 2020, with net profit margin dropping by 20.1%. This translates to an average daily loss of US$230 million. With revenue expected to be just USD$419 billion, airline companies have experienced a 50% revenue decline as compared to 2019 (US$838 billion). Due to the huge changes in the global aviation industry, what changes will take place in the aviation industry?

7.4.1 Major Adjustments in the Global Aviation Landscape

Before the pandemic, Europe, North America, and the Asia Pacific were ranked first, second, and third in air passenger throughput, which accounted for approximately 30%, 28%, and 27% of the global market share. In particular, the influence of China's civil aviation on the world is increasing with the rapid growth of China's civil aviation market. But with the severe impact of COVID-19, there is no doubt that the global aviation industry will face great changes, and this can be seen from its previous experiences.

First, the development of the global civil aviation market is closely related to the rate and quality of economic and social development. When the economy is stable, the civil aviation market grows at a stable rate. When the economy is fluctuating, the fluctuation of the civil aviation market is much larger than the fluctuations of the economy. This is why the civil aviation market is called the "barometer" of the economy.

In 1971, due to the US economic crisis caused by the combination of trade deficit, the collapse of the Bretton Woods system, and the oil crisis, the growth of global civil aviation dropped from 16% in the previous year to 8%. This also ended the period of the rapid growth of global civil aviation. In 1973, the oil crisis caused by the war in the Middle East and the shadows of the world war directly led to the collapse of the aviation industry in 1974 and 1975. After growth of 13.5% in the previous year, global civil aviation fell below 4% for two consecutive years. In 2001, the Internet bubble and the September 11 terrorist attacks in the United States negatively affected global civil aviation, causing a drop from 8% in the previous year to –4%.

The main cause for these events is that the aviation industry has a long industrial chain that involves many industries. In particular, the energy and finance industry has a greater impact on the aviation industry as they, to some extent, are a collection of various industries. At the same time, due to its close international links, the aviation industry shares a direct and exponential relationship with economic growth.

Furthermore, when major events coincide with global economic downturns, it usually takes 2 to 3 years for the aviation industry to recover from the crisis. This period is typically marked by a wave of airline restructuring to cope with operational difficulties, with the September 11 attacks as a classic example.

On 11 September 2001, 19 terrorists brought knives, toxic sprays, and other weapons on board. After killing the pilots and hijacking four civilian planes during flight, the terrorists crashed the planes into the World Trade Center and the Pentagon, causing the deaths of at least 2,996 people on the planes and in the buildings.

This vicious incident has aroused air travel safety concerns. U.S. airspace was closed for three days, and air passenger traffic plummeted. It did not immediately return to normal after airspace was reopened. The U.S. Department of Transportation data showed that U.S. air passenger traffic reached pre-9/11 levels only in July 2004. Even then, passenger load was only 98.3% of its peak in 2001, and the number of employees in the aviation industry decreased by 28%.

From 2002 to 2005, American Airlines, United Airlines, and Delta Airlines filed for bankruptcy protection due to insolvency. After American Airlines filed for bankruptcy protection in 2002, it paid off part of its debts under government guarantees. In 2003, it withdrew from the bankruptcy process. However, it did not last long. In 2004, it filed for bankruptcy protection again. It was not until it acquired American West Airlines in 2005 that things returned to normalcy. United Airlines and Delta Airlines also returned to normalcy after absorbing Continental Airlines and Northwest Airlines, respectively. The process of merger and integration was the recovery period of the aviation industry.

In fact, over the past 20 years, there have been multiple highly contagious pandemics in the world, including the 2003 SARS pandemic in China, avian flu in China in 2006, and H1N1 flu in the United States in 2009. Referencing the impact of past pandemics, once the crisis is over, it is a universal rule for industry figures to return to normal levels. However, COVID-19 has far exceeded any previous infectious disease crisis in terms of transmission speed, scale, and impact, and has caused a global spread, seriously affecting global economic development and social activities.

In the first half of the year, the pandemic's impact on the aviation industry was apparent. American airlines affected by the pandemic were collectively seeking more than $58 billion in government assistance, including government-guaranteed loans, cash subsidies and tax reductions, etc. This will exceed three times the scale of the 9–11 period. Delta Air Lines announced a reduction of flights by 70%, while the European aviation industry's capacity dropped by 80% in March, with some airlines beginning substantial layoffs. Many governments expressed their support for major airlines. In terms of China's industry-wide profits, the aviation industry suffered losses amounting to 74.07 billion *yuan* in the first half of this year, most of which were attributed to airlines.

There is no doubt that the pandemic's impact on the service industry, including the aviation industry, is more profound than any infectious disease. Regardless of whether it triggers a global economic crisis, an aviation industry crisis has already emerged. In terms of civil aviation growth, due to the pandemic's global reach, the impact is likely to last until the third quarter, which is the traditional peak season. Moreover, due to the global economic recession, the recovery speed of travel demand will be slowed down, and it will be difficult for a V-shaped rebound, or one similar to post-SARS, to occur.

IATA predicts that passenger travel volume will drop 38% year-on-year in 2020 as compared to the previous year, with the decline in Europe reaching as much as 46%. This year, the global civil aviation industry may experience the fourth negative growth in nearly 60 years. Based on historical experience, this pandemic will not only have a greater impact on the pace of development and revenue performance of this year but will also result in the industry recovery lasting over the next 1–2 years, as well as major adjustments.

7.4.2 New Aviation Trends after the Pandemic

As a result of the pandemic, although the civil aviation industry has shown obvious signs of recovery, China still faces global aviation challenges and further uncertainties in the post-pandemic era, including the aviation industry's development challenges and structural adjustments, reforms, and innovations in the digital economy, and changes in business models.

Firstly, operations in the first half of 2020 came to a standstill due to the pandemic. However, due to the relatively good pandemic control in China, the domestic aviation market may further recover in the second half of the year. CAPSE released the Analysis of Visitors' Willingness to Travel Index for July 2020, reflecting a significant increase in tourists' willingness to travel during the summer vacation. In July, it increased by 39% compared to June. Additionally, 42% of travelers had travel plans from 1 August to 15 August.

To promote the industry's stable development, the Civil Aviation Administration of China actively communicated and coordinated with relevant departments and introduced 8 support policies based on 16 previously issued support policies. For the financial aspect, it reduced or exempted airlines' civil aviation development funds. Industry-wise, it continued optimizing airline flight licensing management, simplifying flight approval procedures and scheduling coordination procedures, as well as other measures. For cargo, it enhanced support for freedoms of the air and strengthened infrastructure and information to help companies reduce their burden. It is predicted that this "16 + 8" support policy will reduce airlines' burden by about 10 billion *yuan* annually.

Secondly, China's aviation industry will reshape itself this year. Since the country's economic development has entered a steady period in 2019, its civil aviation industry has also shifted from a period of rapid growth to a period of steady growth. In terms of its structure, many small and medium-sized airlines were still in a stage of melee and poor profitability before the pandemic. Barring external incidents, China's domestic aviation industry will adjust by following existing trends and it will gradually ensure the survival of the fittest. Undoubtedly, the outbreak will hit the civil aviation industry, but it will also accelerate the adjustments.

The governments at all levels have taken various measures to help airlines cope with their difficulties during the pandemic. However, in the case of a full-blown aviation industry crisis, some airlines will not survive, resulting in readjustments to China's civil aviation industry structure. In terms of global practice, the best choice when an aviation industry crisis occurs is to first help leading companies recover as soon as possible. They will then absorb and merge with companies, which were unable to operate. This will stabilize the industry.

Thirdly, the pandemic will promote the systematic development of industry reform and innovation. After that, relevant organizations may have a clearer understanding of aviation industry laws, which will help them to gradually manage challenges by promoting service at their core, as well as regulatory upgrades, airspace reforms, and service innovations. Concurrently, the pandemic will undermine the thought of establishing airlines by private capital and local governments. As the industry undergoes adjustments, the allocation of core resources will be optimized, and it will enter a period of orderly competition. As its structure gradually stabilizes, airlines that complete internal integration first will have the ability to expand outwards and strengthen their international competitiveness.

Finally, the pandemic will strengthen the industrial chain's integration and promote the transformation of business models. It has catalyzed the accelerated development of the digital economy, and online commerce is now in trend. However, the service industry as a whole will be hit hard, and this is an opportunity to promote the integration of upstream and downstream related industries in the aviation industry.

Currently, air tickets sold by OTAs generally exceed 30%. As the pandemic ensues, OTA companies struggle to survive. At the same time, while airlines have always accorded great importance to online business, the effect of self-built OTA is unsatisfactory. But the pandemic will bring a historical opportunity for both sides to jointly build an organic community. Airlines can save marketing and customer search costs, while online travel companies can save purchasing costs, resulting in a win-win situation for both parties. On this basis, a larger platform for cooperation among airports, airlines, railway companies, car rental companies, OTAs, and hotels can be established to serve as a comprehensive travel service provider.

From the two World Wars, the oil crisis, September 11 attacks to the 2008 financial crisis, the global aviation industry has experienced several major shocks and showed great resilience. Today's aviation manufacturing industry is more resilient and flexible to resist crises than at any time in history. Leveraging the power of the digital economy, the global aviation industry will gradually reveal new directions and pathways moving forward.

Of course, just like how countries cannot get rid of the economic cycle, companies are unable to get rid of economic challenges, be it large or small. Although this pandemic will accelerate the survival of the fittest among air transport companies in various regions around the world, it will also forge an enhanced track for the aviation industry in the post-pandemic era.

7.5 Hollywood's Refuge and Rise, Darkness and Dawn

As COVID-19 continues its global spread, the entire film and television industry has entered an ice age – all aspects of film and television production have been affected. From the large-scale indefinite closure of theaters, the inability of many producers and investors unable to withdraw funds, stagnant state of film and television production and shooting, to the ongoing unemployment of practitioners, Hollywood, as a global film and television center, is bearing the brunt of the blow.

Among them, major Hollywood studios such as Disney have repeatedly postponed the release time of movies. After postponing the release of Mulan four times in 2020, Disney is ready to hit the screens.

On 5 August 2020, CEO Bob Chapek announced that fans will be able to watch Mulan online, priced at $29.99 (approximately 208 *yuan*) on Disney's streaming platform Disney+

from 4 September. At the same time, Chapek said that Disney will release Mulan in theaters on the same date in countries/regions where Disney+ is not available.

The financial report released with the news of Mulan showed that in the third fiscal quarter, Disney's revenue fell 42% year-on-year, with a net loss of US$4.718 billion, compared with a profit of US$1.43 billion in the same period last year.

The pandemic has dealt a huge blow to major Hollywood film companies such as Disney, and the entire film and television industry has entered an ice age. However, this is not the first time Hollywood has experienced its darkest moment. For the global film and television industry, the opportunity for change may be coming.

7.5.1 The Hollywood Storm

In 1893, Edison invented the movie mirror and created the "Prison Car" photography field, which is regarded as the beginning of the American film history. In 1896, the Weitai projector introduced the mass screening of American films.

The high revenue of movies triggered fierce competition. In 1897, Edison filed a lawsuit over a patent dispute. By 1908, a film patent company, controlled by Edison, was established. It had 16 patents. In 1910, the film patent company almost monopolized the links in production, distribution, and screening of American films at the time, including film production and equipment manufacturing. Any independent producer who wants to share with them had to pay them copyright taxes.

In the beginning, Hollywood Gardens in Southern California were just a shelter to avoid "destroyers" sent by film patent companies to stop them from producing movies. As a result, Southern California has stepped up publicity for its beautiful filming locations and low costs. More importantly, it became a gathering place for independent filmmakers to avoid subpoenas that violated trust licensing procedures. By 1914, independent filmmakers from all over moved to Hollywood. This was the embryonic form of the future Hollywood studio system.

Griffith, the father of American film, joined Bivograf in 1907 and directed the first film The Adventures ofTao Li the following year. By 1912, Griffith had produced nearly 400 films for the company. He gradually shifted the focus of filming to Hollywood, discovered and trained many famous actors, such as Mack Sennett, Mary Pickford, and the Gish sisters.

Just before World War I, nickel cinemas were gradually replaced by cinemas with relatively better conditions. The monopoly power of the film patent company gradually diminished and was removed in 1915. At this time, a group of new film artists represented by Griffith appeared, and the production center moved from East Coast to Hollywood. While the First World War negatively impacted the film industry of European countries to varying degrees, it contributed to the rise of American films, which constantly flooded the European market. By the end of the First World War, hegemony in Europe was established.

In 1926, Warner Bros. released the first sound film Don Juan. In the beginning, Vitawind could not synchronize the sound with the picture. But in 1927, Warner Bros. launched the

Broadway popular song and dance film The Jazz Singer, which resolved the issue. On 6 July 1928, Warner Corporation released another 100% sound film titled *Lights of New York*. Since then, sound films were launched, and audio equipment was installed in half of the American theaters within a year.

By 1930, all feature films were sound films, apart from Chaplin, who continued shooting several silent films. This also ushered in a new era of film production with new media methods, cinema equipment, and interpretation methods.

The sound revolution of movies happened at the same time as the Great Depression of 1929 in the United States. It was also the time when the American film industry ushered in a golden age lasting 20 years. The Rockefeller consortium and the Morgan consortium began a fierce battle of control over Hollywood. By 1935, the two consortiums controlled eight major Hollywood companies, namely Paramount, Fox, MGM, Warner Bros., Thunderbolt, United Art, and Columbia, Universal. They became more financially powerful, each expanding their strengths and recruiting talents, and controlled 95% of the US film distribution.

In the 1920s, the first long-term audiences appeared in movie theaters. In the golden age of the 20 years from 1929 to 1949, an average of 83 million Americans went to the movies every week. After the end of World War II, due to the expansion of the film market, Hollywood once again presented a thriving scene. A group of film masters returned from the battlefields and shot a batch of high-quality and well-loved works. But just as Hollywood was thriving and flourishing, a political storm changed the direction of the American film industry.

In 1947, the House Un-American Activities Committee (HUAC), a committee of the U.S. House of Representatives accused some communists of infiltrating the film industry and using films to promote themselves. The film industry became one of the targets of the committee's investigation. The HUAC began its trial on October 20, 1949. As a result, the "Hollywood Ten" were sentenced to jail, and a large number of directors, producers, actors, and screenwriters were blacklisted and were unemployed or went into exile. Chaplin was one of them. This turmoil, which lasted for more than 10 years, greatly impacted Hollywood's growth.

In addition, in the late 1940s, independent exhibitors took reference from the antitrust law and began protesting to the government that large studios engaged in monopolistic behavior by leveraging advantages of their one-stop production and sales to exclude independent manufacturers. In 1948, the Supreme Court declared that the monopoly of the film industry by major studios was illegal and required the screening industry to be separated from the distribution and production industries. This case called the "Paramount Judgment" marks the end of the classic Hollywood era and Hollywood's major studio system.

The ruling caused major studios to lose most of their first-round theaters, which forced the studio to transform from a vertically integrated monopoly company to a production company focused on distribution. As a result, independent filmmakers were given more opportunities, and Hollywood's traditional economic model began to change. The former big studios started renting studios to independent producers. With the death of the contract star system, the brokerage industry grew rapidly. Those new films seemed ready to welcome the upcoming rejuvenation. However, the challenge was far from over.

In the 1960s, war was spreading in all directions, and Kennedy's democratic policy of focusing on civil rights was supported by students across the country. The civil rights war swept across the country, anti-war waves and radicalism prevailed, and the women's liberation movement was also on the rise. After the war, the offspring of the "postwar baby boom generation" gradually grew up, and the audience's attitudes toward movies underwent unpredictable changes, accompanied by a sharp increase in film production costs.

In fact, after the disintegration of the large studio system, large film companies had no theater chains to guarantee the release of their self-made films. They had to distribute each film separately and rely on the quality of the film to compete for audiences, leading to a significant increase in production and publicity costs.

In the classic Hollywood period, the cost of each film was about US$500,000. By 1952, it reached US$1 million. It was US$1.5 million in 1961. By 1970, the average cost reached US$2 million, doubling in just ten years. However, between 1972 and 1977, the cost jumped 178%, roughly four times the national inflation rate. From 1977 to 1979, it rose from about 7.5 million U.S. dollars to about 10 million U.S. dollars.

The stakes were constantly rising, and these studios could even be described as crazy. In the 20th century, Fox, bankrupted by Cleopatra, was in the black after The Sound of Music, and major companies rushed to replicate it. This was an era in which a movie can save a company. The changing tastes of mainstream audiences made this game even more exciting. However, one of the direct reasons why viewers no longer choose to watch movies in cinemas was the prevalence of television.

The number of TVs increased from 940,000 in 1949 to 3.875 million. Between 1951 and 1953, the number of TVs jumped from 10.3 million to an astounding 20.4 million, and 34.9 million in 1956. By the end of the 1950s, 90% of American households had a TV. During this period, the number of movie viewers also fell significantly. The dark moment had arrived.

The massive loss of audiences forced studios to find ways to improve the audience's viewing experience so that they could return to the theaters. With a huge market demand, a series of technological innovations were realized, for instance, film quality, sound recordings, automatic focus lens and new materials etc.

At the same time, Hollywood was in a transitional period. After the success of the color film The Wizard of Oz, which was previously unpopular in 1939, music film and color became mainstream. However, there remained a missing element amidst the change. After Kodak invented Eastman color film in the 1950s, the number of color films increased significantly. However, less than half of the films used color. What made Hollywood give up black and white films was the popularization of color television sets in the 1960s.

However, the innovation arising from color remained unsatisfactory to some. Some still wanted color films, mostly used for symbolism, to be screened in black and white. Eventually, this transitional period had to be finalized with the widescreen, an invention with far-reaching aesthetic significance.

In general, the application of new technologies in the film industry underwent three stages. It began with the invention, followed by adapting the product to the market. The final stage was the final expansion. For example, the widescreen was invented in the 1920s, but there was not enough profit potential to promote its continued development. For 25 years, it was awaiting an opportunity for its advancement to the second stage. In 1952, This is Cinerama, which was screened on Broadway in New York, was the first work of the film's widescreen revolution and enabled the technology to progress to the second stage.

The early widescreens were dominated by the landscape and factual movies. The movie, How to Marry a Millionaire, resulted in the support of this technology with grander themes. It also motivated major studios and exhibitors to upgrade their equipment. The widescreen seemed to be the antidote for filmmakers to save the box office. The next few films were also successful, thus laying the foundation of the high box office for widescreens.

In fact, before the popularity of the widescreen, American movies used the 1.37:1 academic screen aspect ratio. During the 1950s technological evolution, the ratio became 2.35:1, while TV used the previous narrow screen ratio of 1.33:1. There was no doubt that the 19-meter-long, 8-meter-wide color stereo widescreen movie was significantly better than the 33–48 cm black-and-white TV.

After a series of successes, the top widescreen technology, the Todd-AO system, became the norm.

In 1955, Oklahoma, adapted from the musical and equipped with 70mm projection film and Dolby stereo, was released. Its target audience was mainly the upper class or social elites who went to the movies. To adapt to a more luxurious screening format, the Todd-AO system proposed a screening speed of 30 frames, which were used by just Oklahoma and Around the World in 80 Days. Due to the unobvious effect and inconvenient conversion, it was restored to 24 grids thereafter. Although the innovation in frame rate was not successful, the stereo widescreen was made mainstream by the Todd-AO system.

The continuous innovation of enabled the film industry to remain active in the public's eyes. However, the benefits of large studios remain to be seen with the excessive loss of audiences and increasing interest in entertainment activities such as television. Finally, the studios began cooperating with TV networks as a content provider, and the competition between them ended with a handshake. As the US economy embarked on a period of rapid and stable development, a new dawn has once again resurfaced in Hollywood.

7.5.2 How Did the Pandemic Change Hollywood?

After the 1980s, new cinemas with multi-hall theaters combined with large shopping malls appeared, ushering in a new era for Hollywood. At this stage, Hollywood was no longer limited to only producing movies. Various cultural media industries such as radio and television, cable television, satellite television, music, viewing sports, and theme parks, had integrated with Hollywood. This new model of horizontal integration became popular among Hollywood studios.

Hollywood has once again entered the golden stage of development and this was called the New Hollywood. But it did not bring about another rapid growth for the American film industry. However, starting from 1981, the industry began entering a period of stable development. The film box office was increasing year by year, with a relatively stable growth rate with no major fluctuations.

In 1980, the North American box office was US$2.749 billion, which almost doubled to 5.033 billion US dollars in 1990. Since 1995, there have been about 10 blockbuster movies with a box office revenue of over 100 million *yuan* every year. The trend reflected the strong rising economic strength and national strength of the United States after the end of the Cold War in the 1990s, as well as the beginning of commercialization in Hollywood movies. By 2012, the total box office of North American movies reached 10.811 billion US dollars.

In the late 1980s, the Reagan administration began relaxing corporate control policies, and several major production companies began regaining screen ownership. Since 1986, major companies had purchased or invested in theater chains. After experiencing film industry improvements in the 1980s, in the 1990s Hollywood movies were akin to European football leagues, and it became a smoke-free industry in the United States. Movie screens were also increasingly concentrated in major theaters. Taking the mid-1980s as an example, major cinema chains controlled 35% of the country's 22,000 screens, of which 7,000 premium screens were located in the prosperous big cities and contributed to 80% of the box office revenue.

In addition, the advent of digital technology had a major impact on movies. Firstly, digital technology reduced the cost of film production, especially offline costs such as equipment and technical production costs. However, the cost of online creation, for instance, the cost of celebrities, scripts, and directors remained constant. The status of traditional studios and production centers and the power of production unions had weakened, while the value of existing creative resources had increased, thus somewhat reducing the barriers to market access.

Secondly, digital technology affected the way movies were distributed. The changes affected the two-way high-speed digital communication network and a media industry that moved from a regional geographic monopoly (cable network) or oligopoly (urban newspapers, movie theaters) to a highly competitive media structure. This industry was shaped by language, culture, and lifestyles, as well as the sharply rising value of existing content (films, programs) reserves, especially those that were provided to segmented target markets.

In addition, box office revenue did not account for the bulk of a movie's revenue. In 1999, the box office revenue of American films in the United States and the world only accounted for 26% of its total revenue, while videotapes and television distribution became the main revenue generators. Fourthly, the number of television channels increased after digitalization, and personalized movie service channels such as pay-per-view (PPV) and quasi-video On-demand (NVOD), and video-on-demand (VOD) made viewers pay more to watch movies.

Lastly, digital technology reduced film distribution costs. The mass media supported by traditional advertising was undermined, and the advent of new forms of information circulation reduced transaction costs, resulting in relatively lower creative and media promotion costs. Most importantly, the power of existing communication and entertainment brands or resources with unique identity or the brand value was enhanced. However, as long as Hollywood continues to monopolize distribution, new technologies in production and distribution cannot impact its position.

Hollywood had experienced a rise, a boom, followed by a decline. It is undergoing industry changes to this day. When the pandemic hits the film industry hard, what kind of changes would the film industry be ushering in?

In reality, in the past decade or so, Hollywood movies had once again undergone another round of profound changes in production subjects, product strategies, narrative methods, and communication mechanisms. Film companies represented by Disney, which introduced innovations to the film production paradigm, overseas markets represented by China are further reshaping the Hollywood industry ecology, and emerging digital players represented by Netflix are challenging the existing industrial structure.

Although the overall growth rate of North American movie box office revenue was slow and fluctuated from time to time, beneath the numbers was an indisputable new ecology and reality. As Ben Fritz pointed out, this was "a grand story of a new Hollywood." In this story, franchising and brands dominate, original ideas and celebrities are marginalized, and television and movies have exchanged positions in the culture and economy sectors.

In recent years, the evolution of the Hollywood film industry model has not developed in silos. It is the result of intertwined forces in the global entertainment industry. Hollywood's evolution, the development of other entertainment industries such as television, new changes in the global film market, the rapid advancement of digitalization and media integration, and many other factors combined had given rise to the birth of the current industrial and cultural Hollywood movies landscape.

In early 2019, streaming service provider Netflix officially joined the Motion Picture Association of America (MPAA), becoming the first non-Hollywood film studio to join it. On 20 March 2019, U.S. time, The Walt Disney Company officially completed its transaction with 21st Century Fox, acquiring most of the latter's assets for $71.3 billion. After this transaction, the history of six major Hollywood studios has officially ended.

This also showed that the subversion of Hollywood by Netflix and others had begun. High-quality streaming media content directly impacted the interest of traditional movie theaters, causing conservatives in Hollywood to resist streaming media created by technology giants such as Netflix and Amazon, which was evident in director Spielberg's proposal to Oscar in 2019 to cancel Netflix's Oscar qualification.

Of course, this proposal was eventually rejected by the American Academy of Motion Picture Arts and Sciences, and there were multiple reasons behind that decision. Perhaps, the most direct one was the warning by the US Department of Justice's antitrust department to the Oscars. This was because Oscar's rejection of streaming movies might violate the antitrust law. Optimists in the industry believed that the pandemic had accelerated the integration of Hollywood and streaming media because they are not directly opposing each other.

However, it is undeniable that film technology innovation will continue to be pushed to the forefront as the market changes, and the economic structure of modern society will continue to diversify. The impact of the digital information age has caused countless industries at their peaks to decline and vice-versa, the technological evolution in the age of technology has also become faster. In the future, the level of cooperation between the film industry and streaming media may be higher, akin to the significance of silent films to sound films in the industry, or the huge changes brought about by 3D technology. The opportunity for change may already be here.

7.6 The Pandemic Shapes a New Paradigm for the Digital Economy

In the short term, the pandemic will undoubtedly worsen the global economy. The global supply chain, employment, livelihood, business survival, policy response, and even social ideology are facing major challenges. From a longer-term historical perspective, COVID-19 has caused a unique situation where both supply and demand are weak, while it has also accelerated the challenging transformation of the global economy.

In recent years, the digital economy has continued rising in popularity, and it became a keyword in China's two parliamentary sessions in 2020. In fact, we are moving from the physical to the digital world. Although the popularization of computers, smartphones, and network technologies have initially changed people's lives and consumption habits, the digital economy is not purely an industrial revolution. It has revolutionized value creation and redefined the process of value distribution, which were substantially different from the old ideas, old order, and old classes rooted in the traditional economy. Therefore, the integration and unification of the digital economy and new ideological order are still in the process of integration, while continue to reverse the evolution of the digital economy.

As a catalyst, the pandemic may accelerate the formation of social order and moral culture adapted to the digital economy. It may also coerce us to completely detach ourselves from the physical world for the first time and reflect on the causes and effects within. The true digital revolution is coming.

7.6.1 The Digital Transformation of Enterprises Spurred by the Pandemic

The pandemic's impact is a "stability test" for companies. In this test, many companies found it is difficult for the traditional production methods and sales methods are to quickly respond to the rapid changes in the industry and supply chain. Therefore, they will accelerate their search for changes and breakthroughs in production methods and sales methods and strengthen the situation in the face of uncertainty to maintain the company's survival and gain a lasting competitive advantage in the market. Through the survey of companies during the pandemic, it was found that the more digitalized companies were less impacted by the epidemic. Digital transformation has become a key strategy for companies to deal with external uncertainties.

In terms of production methods, companies developed a more comprehensive and profound understanding of the availability, ease of use, and usefulness of digital technologies such as artificial intelligence, big data, and cloud computing during the pandemic. This understanding has reduced barriers to technological cognition and will further accelerate the widespread and in-depth application of digital technology. In the post-pandemic era, the digital transformation of enterprises will be promoted by the following aspects. Firstly, enterprises will accelerate the adoption of intelligent analysis tools such as artificial intelligence and big data in business competition, and will accurately predict industry changes and market trends based on leading indicators, which will serve as the basis for decision-making and inventory management. The second is to further accelerate the digital, networked, and intelligent transformation of traditional production equipment, introduce intelligent production lines on a larger scale, and meet the elastic changes of market demand within a shorter timeframe. Thirdly, the speed of enterprises onboard cloud servers has been further accelerated, and cost structure adjustments can be made more flexibly through cloud migration to reduce variable and fixed costs.

In addition to changes in production methods, the pandemic will catalyze the transformation of enterprises' digital sales. Traditional sales are mostly offline and were greatly impacted by geographical isolation during the pandemic. To hedge the losses caused by the pandemic, traditional manufacturing leaders such as SANY Group and Gree Electric in mainland China have tried online live broadcasts and short-form video marketing. Gree's Dong Mingzhu conducted live stream shopping broadcasts thrice in three weeks. Although the first session only achieved 225,00 *yuan* of sales, the second and third sessions achieved sales of 310 million and 703 million *yuan* respectively. SANY Group also held three live broadcasts during the pandemic, covering factories, Spring Festival, and shopping festivals.

During the live broadcast of a shopping festival, SANY Group, which answered product queries online and advanced its promotion mechanisms, achieved 288 sales in two hours, completed 186 3,000-*yuan* large orders, and amassed sales of 50 million *yuan*. The success of Gree and SANY Group has proven the feasibility of digital sales. In the future, more traditional enterprises will try digital sales transformation.

From a business management paradigm perspective, there will be varying degrees of online transfer trends throughout the process. Due to the difficulties of resuming work and production, most companies realized the shortcomings of traditional offline management in combating risks, so they began promoting remote management to catalyze the transformation of the business management paradigm in the digital economy era. The process of business management and value chains have seen different levels of online transfer trends, especially the rapid development of remote offices and cloud signing (electronic contracts), while remote management of suppliers and customers are also gaining high levels of growth.

Coase's theory of enterprises and markets believes that enterprises appear because they can reduce transaction costs when leveraging the market for resource allocation and improve resource allocation efficiency through administrative instructions from within. The essence of a remote office is to reduce the transaction costs of enterprises, optimize the internal management mechanism within and strengthen interactions with the market through cloud signing (electronic contracts) beyond the office. Also, companies can strengthen cross-regional global collaboration through remote management without having to pay high transaction costs.

Therefore, although on-site management will continue to dominate for a long time in the future, the proportion of remote management in corporate management, as an effective supplement to promote corporate management efficiency, will significantly increase. This can be attributed to the enhancement of corporate digitalization and the strengthening of remote management awareness.

Companies will use digital technology to predict and simulate various impact events to speed up the development of remote management programs that can improve their agile response to crises. In addition, the transformation of the business management paradigm will increase the demand for remote management platforms and software and promote the explosive growth of the industry. Enterprises will explore more digital transformation of management services, thereby promoting the migration of management services and integration of business data and management data on the cloud.

In addition, digital construction, telemedicine, and smart agriculture have become important directions for the digital transformation of traditional enterprises. According to IDC forecasts, more than 25% of global GDP output in 2020 will be driven by numbers. At the same time, it is expected that the global market's digital-driven GDP will account for more than 50% by 2023.

7.6.2 New Styles of Digital Consumption

During the pandemic, a large number of consumer behaviors shifted from offline to online purchases, which promoted the transformation of business models in the consumer sector and accelerated the emergence of new formats and models of digital consumption.

The rise of digital consumption is in line with the general economic logic. Consumers first enhanced their immersive experience of new business formats, new models, and new applications during the epidemic, and their preference and stickiness for websites and applications were enhanced. Subsequently, the epidemic forced consumers to actively complete consumer education. Through a long period of concentrated learning behavior, the operational skills mastered by consumers become stock skills. This has broken the "high-cost wall" of market promotion of new business formats, new models, and new applications, and the user penetration rate has been systematically improved. It is worth mentioning that digital consumption and traditional consumption are not completely substituted. Digital consumption can further expand the demand by further mining the potential needs of consumers and have a leading role in traditional consumption.

On the one hand, there are increasing new models of relatively mature formats, including online group buying, intelligent logistics and distribution, and fresh food. Taking China as an example, the demand for home-to-home services in supermarkets such as fresh food has risen sharply due to a falling number of consumers going out during the pandemic. The number of active users of the main fresh-to-home platforms, the number of daily usages per capita, and the length of time have increased significantly. As far as Hema Xiansheng is concerned, the contribution of online purchases in Q1 2020 will account for approximately 60% of Hema's GMV, an increase of 10% year-on-year.

The US e-commerce giant Amazon, which saw global sales soaring to US$11,000 per second during the pandemic, benefited from the closure of many physical stores. According to Amazon's Q1 2020 report, North America, comprising mainly consumer and online retail sales, recorded a revenue of US$46.127 billion, which was a year-on-year increase of 28.8%, as compared to US$35.812 billion in the same period last year. The international portion mainly includes revenue from consumer goods and online orders from the international market, and it includes export sales from these key international online stores (excluding export sales from North American online stores), with a revenue of 19.106 billion U.S. dollars. This is an increase of 18% year-on-year as compared with 16.192 billion in the same period last year.

On the other hand, there is a rapid increase of new models, such as online education, Internet healthcare, cloud entertainment, and cloud tourism, in the development stage. Among them, short videos leverage the sudden increase in 4G traffic and firmly establish the dual-power structure. Backed by the rapid increase in mobile Internet traffic in recent years, short video applications have become a big thing within a few years, changing the competitive landscape of the Internet industry. According to Questmobile's statistics, short video applications accounted for only 4% of China's Internet usage in Q1 2017. By Q3 2019, it surged to 19%. In September 2019, short videos accounted for 64% of the increase in user monthly usage time, far exceeding other application types.

Among them, the number of users of Douyin and Kuaishou exceeds other short video applications. The dual-power structure is firmly established, with Douyin enjoying a relatively quicker growth momentum. As of April 2020, Sensor Tower store intelligence data showed that the total number of downloads of ByteDance's Douyin and its overseas version TikTok, in the global App Store and Google Play has exceeded 2 billion. After reaching 1.5 billion downloads, the application only took 5 months to achieve this new milestone. In the first quarter of 2020, Douyin and the overseas version of TikTok received a total of 315 million downloads in the global App Store and Google Play, making it the world's most downloaded mobile application.

For educational live streaming, the pandemic has resulted in students and institutions canceling offline lessons, and live broadcasts became the best alternative for these courses. Due to the large number of students and resumption of classes, live-streaming for education has sprung up. For example, Douyin, which is based on educational resources accumulated by the DOU Knowledge Plan and its in-house Gogokid for K12 teaching and Open Language, saw its education live broadcasts rapidly increase in February. The number of anchors, sessions and viewers increased 110%, 200%, and 550% respectively on a month-on-month basis. According to Frost & Sullivan's survey, the market for K12 online tutoring in 2019 is 64 billion *yuan*, with an online penetration rate of 15.7%. At the same time, Frost & Sullivan predicts that the market size of K-12 online tutoring is expected to grow to 367.2 billion *yuan* by 2023, and the online penetration rate to hit 45%. Based on the above forecast, the compound growth of China's K12 online tutoring market will reach 55% in 2019–2023.

At present, the development of digital consumption will present at least three changes. First, the increase in the number of offline small, medium, and micro enterprises exiting, while large enterprises accelerate their offline business integration. The market concentration will continue to increase, and the industrial organization structure will continue to optimize. Secondly, users will further develop online consumption as a habit, while the scale of the digital economy in its mature period will continue to expand, and the new digital economy in the growth period will continue to quickly explode. Thirdly, accelerating online and offline integration will be a long-term trend of economic development. Offline companies will not be completely subverted. The provision of personalized, differentiated, and high-quality services will become an important competitive strategy.

According to the National Bureau of Statistics of China (NBS) data from January to February 2020, the value-added of industrial enterprises above the designated size decreased by 13.5% year on year, while smartwatches and bracelets bucked the trend by increasing 119.7% and 45.15% respectively. The service industry production index dropped by 13.0% year on year, while the information transmission, software, and information technology service industries achieved an increase of 3.8%. The total retail sales of consumer goods decreased by 20.5% year-on-year, while the online retail sales of physical goods increased by 3.0% year-on-year. From this series of comparative data, it can be seen that traditional industries representing old drivers have shown clear powerlessness in dealing with exogenous shocks, while the new drivers represented by the digital economy have shown great development potential in hedging uncertainties. The ability of digital technology to enhance an economy's flexibility has been fully reflected.

Therefore, the pandemic has activated the fast-forward button for a new wave of technological and industrial changes, injecting greater flexibility into the economic system. Enabling the digital economy as a macroeconomic stabilizer, buffer, and accelerator, so that the national economy can adjust production, distribution, and consumption more resiliently in the face of shocks has become the consensus and direction of the next economic development.

Concurrently, it is necessary to further accelerate the construction of new infrastructure such as 5G deployment and artificial intelligence, development of the digital economy and industrial Internet, digitalization, networking, and the intelligent transformation of traditional industries, as well as the replacement of old drivers with new ones.

Looking back, we have to concede that the pandemic is guiding the direction in pain. Looking at the major plagues which changed the trajectory of human history, for instance, ancient Greece lost its glory due to the Athens plague, while the originally difficult social transformation in Europe in the Middle Ages became seamless due to the "Black Death" and laid the foundation for the Renaissance, English Reformation and Age of Enlightenment periods. Today, medical technology, rescue facilities, and isolation measures have undergone fundamental changes. Human beings are no longer ignorant about the virus, but the global spread of COVID-19 is still affecting the trajectory of today's society.

From a long-term perspective, while the sudden and unexpected pandemic has revealed the illusions of prosperity, it has brought about digital transformation without the heavy shackles and accelerated the formation of a long-term trend. The trajectory of our society may just change accordingly. The pandemic is akin to pressing a fast-forward button. A digital revolution that intends to reshape the global economic structure is gaining momentum, allowing the global economy to break the zero-sum game, thereby forging growth and creating a new track for greater competition.

8

Great Divergence of the Pandemic

8.1 Four Models of a Global Fight Against the Pandemic

The global COVID-19 pandemic has not only exerted huge pressure on countries, including severe challenges to public health and safety, but also on its comprehensive strength, national governance systems and capabilities, and national systems, resulting in greater uncertainty for the future. These made countries work harder to explore anti-pandemic models which were customized following national conditions to reduce uncertainty and form a new production and life order.

The degree of control of the pandemic in various countries can be roughly divided into four models, namely China's strict isolation, Sweden's herd immunity, a mixed-mode model, and the United States' "free fight against the pandemic."

Mixed mode is a choice for many countries. However, it is applied only when the same choice corresponds to each distinctive mode, and so different results have also appeared. However, whether it is to take strict precautions or "lie-flat," or somewhere in between, all countries were forced to make difficult choices due to the lack of ideal means to fight against new pathogens. Against this backdrop, the United States' "free fight against the pandemic" has turned the crisis into a tragedy, and its failure to fight the pandemic is shocking.

8.1.1 China's Strict Guard Against the Pandemic

China uses a classic isolation model with strict prevention measures.

On 7 June 2020, the White Paper China's Action to Fight the COVID-19 Pandemic issued by the Information Office of the State Council of China pointed out that "China has adopted the most comprehensive, strictest and most thorough prevention and control measures. It has adopted unprecedented large-scale isolation measures. We will mobilize resources across the country to carry out large-scale medical treatment, not to miss any infected person, not to give up on every patient, and to admit, treat, test, and isolate whenever required.

The strictly guarded isolation model primarily includes several measures.

The first is to accurately carry out prevention and control work by zones and classifications, and adopt measures such as closing cities, buildings, and communities. In particular, Wuhan's city bus, subway, ferry, and long-distance passenger transport services were suspended, and the airport and railway stations were temporarily closed from 23 January. From 10 February, communities in Hubei province implemented closed management, saving precious time to curb the spread of the pandemic.

For high, medium, and low pandemic prevention and control risk areas outside Hubei Province, Xi Jinping further specified the policy of "categorized guidance" for pandemic prevention and control on 23 February, including emphasizing internal non-proliferation and external defense for Wuhan and Hubei. He also specified the coordination of the adoption of targeted measures in Beijing, Zhejiang, Guangdong, and other provinces and cities with large populations and the surrounding provinces and cities of Hubei, which had to respond flexibly.

The second point was to adopt strict social distancing, including reducing public gatherings and conducting checks on inter-city migrants. Regarding measures to reduce public gatherings, the General Office of the State Council issued a notice on 27 January to extend the Spring Festival holiday and postpone the reopening of all schools. Many large-scale mass gathering activities were either reduced or canceled, while non-essential places attracting crowds such as movie theaters, KTV clubs, and gymnasiums were all closed. Places providing public services such as supermarkets and farmer's markets adopted measures such as entry and exit checks, regular disinfection, and crowd restrictions to effectively prevent the pandemic spread.

To check inter-city persons in mobility, pandemic prevention checkpoints were set up at airports, railway stations, highway intersections, and other major traffic stations.

Pandemic prevention and control agencies in various regions have strengthened guidance on inspection points, implemented strict quarantine measures such as passenger temperature screening, and health and contact information registration. Once a case or suspected case of COVID-19 was discovered, it was to be reported immediately to the local health authorities. Tourists were also persuaded to cooperate with testing.

The third was to implement grid management at the grassroots level by consolidating the responsibilities of territories, departments, units, and individuals for the prevention and control of COVID-19. The pandemic prevention and control teams, established by the masses and led by grassroots leaders, were responsible for grassroots pandemic prevention and control propaganda as well as for the personnel health screenings and mobility, and daily necessities purchases for residents during the community (village) lockdown. The establishment of an "early detection and early report" mechanism was the creation of conditions for "early isolation and early treatment."

In addition, since the pandemic started in Wuhan, there have been at least four localized pandemics in China, namely Harbin, Mudanjiang, Beijing, and Urumqi. However, based on the "zero tolerance" COVID-19 strategy, the city had to remain closed once unlinked (non-imported) cases occurred, and a general test (COVID-19 nucleic acid test) had to be implemented to specifically manage repeated epidemics. It takes around three to four weeks to return the incidence rate to zero and reopen after two weeks.

For example, Beijing authorities announced a new local confirmed case after 56 continuous days of zero reports, on 11 June 2020. As of 16 June, 0000 hours, the number of cases soared. It was reported that 137 new local cases and at least 12 cases of asymptomatic infection were recorded, with many places reclassified as high-risk areas. The resurgence of the pandemic in Beijing aroused great concern across the country. Subsequently, Beijing conducted more than 10 million nucleic acid tests within a week. Beijing's efficiency resulting in the incidence rate falling to zero within a few weeks, once again achieving the "China Speed" of excellent anti-epidemic performance.

China's anti-pandemic model requires huge economic costs because a general survey can easily involve tens of millions of people and incurs costs as high as one billion. At the same time, strong organizational skills are needed, including mobilizing people to participate in testing and organizing tight lockdown measures. It is this mode of strict prevention and defense that the white paper declared to the world that China has won a great victory in the fight against the pandemic. Better epidemic management has also brought about better economic performance. Compared with other countries and regions in the world, China is the only one to announce that its gross domestic product (GDP) in the second quarter was higher than the end of 2019, with Vietnam, South Korea, and Hong Kong trailing behind.

Spain and India's GDP decreased by 20% and 25% in the second quarter as compared to the previous year. In East Asian countries which were able to avoid lockdowns, the impact on the service and construction industries was limited. China's activity data in August showed that as consumers finally re-engaged in the economy, the recovery scope was expanding and achieved without direct government support, unlike Western consumers.

Of course, strict prevention and control measures cannot be achieved without the support of the socialist system with Chinese characteristics. In fact, it is precisely these significant advantages of the "China System" that have transformed into the governance efficiency of the "China Rule," enabling domestic pandemic prevention and control to achieve significant phased results.

8.1.2 The Zen Sweden

If China is the extreme of isolation that strictly guards against death, Sweden is the other extreme that did not close or isolate any city. While travel has been restricted for most, life goes on for the Swedes. Some German scholars said that the "Swedish Route" seemed to be the "comparison group" in this vast experiment.

On 12 March, the Swedish authorities decided to no longer conduct large-scale testing and only test patients admitted to the hospital. There were no "suspected cases" in its published statistics, no calls for the population to wear masks, and no implementation of mandatory social distancing orders. Borders, kindergartens, junior high schools, elementary schools, bars, restaurants, parks, and shops remained open.

However, the population is still required to work from home as much as possible, avoid unnecessary travel, and pay attention to social distancing in public. In addition, the government banned gatherings of more than 50 people, closed museums, canceled sporting events, and built new sanitation facilities.

Obviously, unlike China's strict defense, Sweden has chosen to allow the virus to spread. At the same time, the elderly and vulnerable groups had to be protected to avoid overcrowding in hospitals. As the Swedish anti-pandemic model is unique and individualized, the world's perspective of Sweden is divided into two extreme ends. Some praise it as a model of wisdom, while others denounce it as a humanitarian disaster.

Just as the effectiveness of some severe containment policies is dependent on the national characteristics, the Swedish model is also in line with its characteristics.

On the one hand, Sweden's anti-pandemic model benefits from its natural "social distancing." According to a BBC report, more than half of Sweden's residents lived alone in 2019, among which as many as one-fifth of young people aged 18–25. Compared to Southern Europeans, the Swedes are less enthusiastic about social interactions, and there is an inclination towards a "social anxiety" in Swedish culture. There is a joke in Sweden, which says that the door's peephole should be used to observe the neighbors leave the house, to avoid exchanging pleasantries in the corridor.

On the other hand, the Swedes usually respect the rules and the government. Compared with many other countries, Sweden has not experienced any minor and major scandals and political instability, as well not experienced major natural disasters and wars in the past 200 years. The peaceful scene over the years has made the Swedish society relatively simple and peaceful, resulting in the people's added trust in the government.

The Swedish are split in terms of supporting and opposing the government's handling of the pandemic. However, the Swedish people still chose to trust their government during the crisis. Among them, the polls in the first week of April 2020 showed that two-thirds of the people expressed support for the official measures, including the elderly and the young and middle-aged. Only one-third did not really support it, and most were active on social media such as Facebook and Twitter.

In addition, Sweden is a constitutional monarchy, and its government is more of a service organization. Since the cabinet government cannot directly issue orders to public health institutions, the voices of public health experts are heard, resulting in a greater alignment of Sweden's strategy with science.

For example, the public health bureau first established its modeling based on Chinese data to estimate the approximate number of intensive care (ICU) beds needed. Hence, when the infection began expanding sharply after February, Swedish hospitals were able to respond actively in all aspects. For example, Sweden's largest Karolinska University Hospital increased the number of ICU beds to about 200, five times the original number, in a short period, and the postoperative observation room was quickly converted into an ICU ward. Common beds for relatively mild patients who do not require ICU levels and part of the ward floors were designated as special wards for COVID-19. This is also an important reason why medical resources were not strained in Sweden.

As a result, the natural "social distancing" and the geographical reality of sparsely populated land, coupled with the Swedes who generally respect the rules and the government, have allowed Sweden to embark on a unique Swedish path in its fight against the pandemic.

Of course, Sweden's "lie low" strategy for COVID-19 inevitably brings about damage. At the height of the pandemic in Spring, Sweden experienced a mortality rate far exceeding neighboring countries and this period lasted longer. In particular, the number of deaths in nursing homes was so alarming that it increased the total number of deaths in Sweden. As of 14 May, Sweden's 10 million people reported 3,529 deaths, while the total population of Denmark, Norway, and Finland amounting to more than 16.5 million people, had reported slightly more than 1,000 deaths.

However, the number of COVID-19 cases and deaths in Sweden began to decline over time. At the beginning of September, Sweden had a positive test rate of 1.2%, while the rate at worst-hit Northwestern England was 7%. At the same time, the per capita incidence rate in Sweden was much lower than that of nearby Denmark or the Netherlands.

The society had mixed reviews on Sweden's anti-pandemic model. However, how can we evaluate its needs based on the object of comparison and specific situation? Obviously, the Swedish model does not apply to large countries and is determined by its special national conditions. Compared with East Asia, it cannot be said that the Swedish model is a successful one, especially the large-scale deaths caused by the early outbreaks of large-scale cluster infections in the elderly homes in Sweden, but it is similar to the United Kingdom, France, Italy, and the United Kingdom, which have implemented strict lockdown measures. Compared with Belgium and Spain, the number of deaths per million in Sweden is still far lower than in these countries.

Different models have different mortality rates, and it will be the work of future scientists to judge whether the Swedish experiment was successful.

8.1.3 Mixed Model of Restricted Prevention and Control

In addition to the strictly guarded Chinese model and Sweden's model of herd immunity, there is an anti-pandemic model, which is a mixed model of limited prevention and control. Mixed models were used in countries such as South Korea, Singapore, Japan, India, Russia, Spain, Italy, Germany, and France, etc. In the face of COVID-19, these countries have adopted certain prevention and control measures. But on the other hand, to avoid social panic and to not affect economic development, these countries only attempted to carefully find a balance between "controlling the pandemic" and "protecting the economy" and strived to minimize economic and social costs. Although this anti-pandemic model partially achieved the goal of "protecting the economy," it also made the pandemic last longer.

Among them, South Korea, Singapore, and Japan, which are part of Asia, have mostly learned from China's prevention and control experience of "early detection, early isolation, and early treatment." Although these countries have not adopted strict control measures to "protect the economy," they have successfully controlled the spread of the epidemic with the effective implementation of limited prevention and control measures.

Due to the influence of political systems, values, social culture, and other factors, Russia, Spain, Italy, Germany, France, and other countries always seek a balance between economy and pandemic prevention, isolation and freedom, central and local governments, and a balance among different interest groups. Although China has regularly and proactively notified the World Health Organization, relevant countries, and regional organizations of the pandemic situation from 3 January 2020, alongside the World Health Organization which listed COVID-19 as a "public health emergency of international concern" on 30 January 2020, the above-mentioned countries did not pay enough attention to the warning during the early stages of the pandemic.

Even Italy, which announced the country's "lockdown" earlier, did not implement strict local prevention and control measures. Its "lockdown" measures were mainly aimed at people entering the country from other countries, while Russia ignored the transmission channels of the European pandemic. The returning Russians did not take effective control measures, which led to the worsening of the epidemic. As of 12 June 2020, the cumulative number of confirmed COVID-19 cases in Russia, Spain, Italy, Germany, and France ranks 3rd, 6th, 7th, 9th, and 10th globally respectively.

Undependable Korean lockdown Model

On 19 February 2020, an outbreak occurred at South Korea's "Sincheonji Church of Jesus" outbreak, and Daegu and Gyeongsang became "disaster zones." In the most severe days of the outbreak, South Korea was considered to be the "country with the worst outbreak outside of China." South Korean President Moon Jae-in's approval ratings also dropped to their lowest at that time.

In the face of an outbreak where the number of infected people soared and exceeded 1,000, the South Korean government undertook strong measures after a quick assessment. Through extensive screening, rigorous tracking, and comprehensive treatment, the government not only suppressed the widespread spread of the virus and community infections within three weeks but also avoided a nationwide and regional lockdown, thereby avoiding economic stagnation. This hybrid anti-pandemic model was also considered by the international mainstream media as a successful example of a democratic and free system.

Specifically, in terms of extensive screening, the South Korean government, helmed by Moon Jae-in, upheld the thought of "early detection and early treatment" and announced that South Korea will significantly reduce virus testing requirements as long as doctors classify patients as suspected cases beginning 7 February. There was no need for travel history to a pandemic epidemic area. After mid-February, anyone with suspected symptoms can be tested at any time. As for high-risk groups, mandatory testing measures will be taken.

After the "Sincheonji Church of Jesus" infection cluster, the officials conducted a one-on-one investigation on its believers and required everyone to be tested. In March, after a call center in Seoul and a church in Gyeonggi Province reported large-scale mass infections one after another, the authorities similarly required all close contacts to be tested for the virus as soon as possible.

According to the data provided by the Korean Central Anti-epidemic Countermeasures Headquarters, its virus detection capacity reached 10,000 people just a day after the outbreak in Korea on February 19, and the average daily detection rate was about 15,000 people, reaching a peak of nearly 20,000 people. As of 19 March, 307,000 people were tested for the virus, which meant that 1 out of every 170 people in the country was tested.

In terms of rigorous tracking, suspected and mildly ill patients were required to download the designated mobile app to register their personal information and disclose their geography as mandated by the government. At the same time, various local governments in South Korea were sending the pandemic status to residents. When someone in close proximity was diagnosed, mobile phones around the area will receive an informational warning. Each confirmed patient corresponds to a number, and the travel history of the confirmed patient is disclosed on the premise of protecting the personal information of the confirmed patient.

In terms of the comprehensiveness of treatment, South Korea took measures to send the infected and quarantined to hospital for treatment during the onset of the pandemic. As the numbers were rapidly increasing, South Korea's medical resources became seriously inadequate. This resulted in the death of at least two patients awaiting hospital beds in the worst-hit Daegu City.

Subsequently, South Korea leveraged its strengths and circumvented its weaknesses, adjusted its coping strategies, and switched to "graded treatment," which meant that patients were classified into four levels, namely mild, moderate, severe, and most severe. For those with mild symptoms, they had to receive treatment at home or be admitted to a life treatment center, and those with severe symptoms had to be referred to an isolation ward or an official medical institution designated by the state.

It is worth mentioning that the South Korean government did not implement intensive measures to "maintain social distancing." Instead, it investigated the sources of early cases located the source of cases and places with multiple infections. For example, during an epidemiological investigation in South Korea, it was discovered that nightlife venues, religious gatherings, and indoor gymnasiums were the three main sources of mass infection. Therefore, although there was no "lockdown," the government was explicit in informing the population that they had to be vigilant when going to these three places, requiring everyone to stay at home as much as possible, as well as to cancel or postpone unnecessary travel and social gatherings.

When the pandemic broke out in South Korea, it was successfully controlled using large-scale testing and rapid quarantine measures. According to the Korean Disease Management Headquarters data, as of 20 August, 0000 hours, Korea had a total of 16,346 confirmed COVID-19 cases, of which 14,063 were cured and discharged, and 307 died, registering a mortality rate of about 1.88%. Regarding relevant data released by various countries, the effectiveness of the Korean public health crisis management system in fighting the pandemic was significantly better than that of major developed countries such as the United States, Britain, France, Germany, Italy, Canada, and Japan.

In the absence of mandatory measures to "lock cities or a country," South Korea's economy and society maintained a relatively stable operation. World Health Organization Director-General Tedros Adhanom Ghebreyesus also affirmed South Korea's anti-pandemic efforts on 16 March and believed that it was worth emulating.

The progressive Singapore model

On 23 January 2020, Singapore confirmed its first COVID-19 case. During the early stage of the pandemic, Singapore's epidemic control policy embodied the characteristics of "strictly preventing externally and maintaining normal activity internally." For example, from 1 February 2020, travelers who held a Chinese passport or have been to China in the past 14 days were prohibited from entering or transiting in Singapore. Singapore citizens, permanent residents, and long-term pass holders who had traveled to China had to undergo a 14-day home quarantine after entering the country.

While strictly preventing imported cases, the Singaporean government also strengthened its pandemic prevention measures within the borders. For health protection, it promptly provided people with knowledge about COVID-19 prevention and public health travel advice and raised the coronavirus outbreak alert to orange. It also strictly enforced laws, implemented mandatory quarantine orders for incoming travelers, with those violating Leave of Absence and Quarantine Orders facing up to half a year's imprisonment under the Infectious Disease Act, while foreigners faced penalties such as visa cancellations. Strict measures were also in place to prevent local transmission, alongside increased traffic and maritime traffic inspections on the mainland, and classification of the risk population. It also put forth pandemic prevention measures for preschool children, the elderly, overseas students, and foreign domestic helpers, distributed medical surgical masks to the public, and suspended religious activities to minimize the risk of community transmission.

At the same time, the Singapore government introduced a series of economic support policy measures to protect the people's livelihood. Singaporean citizens, permanent residents, and work permit holders, who traveled to the pandemic areas on or before 31 January and received a Stay Home Notice could apply for a Quarantine Order Allowance (QOA) of SGD 100 (approximately RMB 500) per person per day. The government also met the short-term cash flow needs of companies, helped companies retain and train employees, including providing partial support for employees' wage costs. The Singapore Tourism Board paid 50% of professional cleaning fees for hotels with confirmed and suspected cases. During this period, the livelihoods of taxi drivers (Comfort, CityCab, SMRT Taxis, Tans-Cab, Premier Taxis, Prime, and HDT Taxi) and private hire cars (Grab, Gojek) were greatly affected. In response, the government launched an SGD 77 million Point-To-Point Support Package to help them tide over the difficulties, including a one-time allowance and a daily allowance for each vehicle.

Compared to China's iron-fisted measures, the "prevention" and "screening" by the Singaporean government during the early stages of the pandemic were regarded as a form of "Zen" prevention. By March 2020, the pandemic was raging globally, which spread rapidly in European and American countries and caused a large-scale increase in imported cases in Singapore. It was then that the Singaporean government began adopting more stringent anti-pandemic policies.

In the second stage of Singapore's fight against the pandemic, 28 government departments issued a total of 69 documented measures. During this period, the government's prevention and control measures gradually changed from large-scale to more detailed, technology-intensive, and higher quality.

Firstly, it leveraged technological means to improve the quality of tracking. GovTech and the Ministry of Health jointly released a mobile application called "Trace Together" to support the tracking of confirmed cases, assist in the formation of a tracking chain, and greatly improve tracking efficiency and accuracy.

Secondly, it established social norms more rigorously and carefully to effectively control community cases. The government stipulated work and social norms, disinfection and cleaning measures, and various suggestions for public departments, schools, co-curricular interest classes, financial institutions, religious gatherings, foreign domestic helpers, and other groups of people, including remote meetings, learning in groups, physical lessons once per week, cancellation of foreign domestic helpers' off-days, and temporary suspension of gatherings involving more than 10 people, etc.

Thirdly, it promptly elevated its prevention and control measures. In early April 2020, the confirmed cases of foreign workers in Singapore reflected an explosive trend. On 3 April, Singapore's Prime Minister Lee Hsien Loong announced that starting from 7 April, Singapore will implement the "Circuit Breaker" measures for four weeks.

The measures included closing all non-essential services and retaining only essential public services. All public and private emergency hospitals (including outpatient specialist clinics and surgery centers), community hospitals, general hospitals, public health preparation clinics, and renal dialysis centers continue operating, while non-general practitioner clinics, specialist clinics, dental clinics, and Chinese Traditional Medicine clinics can only provide basic services. Home community care services, such as nursing homes, psychiatric rehabilitation homes, psychiatric shelters, home medical care, home care, and temporary care services will remain open, while elderly care centers, day rehabilitation centers, mental rehabilitation centers, and day hospice care centers were closed. Educational institutions across the country, such as universities, middle schools, and primary schools were converted to home-based learning. Parks and stadiums continued operating, while swimming pools, indoor gymnasiums, gyms, studios, community center courses, museums, libraries, and art galleries and performance venues were closed.

However, the "circuit breaker" measures did not immediately curb the spread of the pandemic. Many new local cases, especially mass infections in dormitories, caused the number of daily confirmed cases to exceed 500 in mid-April 2020. Singapore originally planned to implement the "circuit breaker" measure from April 7 to 4 May. On 21 April, Prime Minister Lee Hsien Loong announced that the measures will be extended to June 1, and Phase 2 will begin on 19 June. This was also the third stage, the strict control stage, of Singapore's anti-pandemic efforts.

The rebound in cases in Singapore in early April was mainly due to the cluster of foreign workers' infections. Therefore, in the third stage of the fight against the pandemic, the government separated the foreign labor groups from the local community groups to block the spread of the virus by placing the workers under quarantine in the dormitories and improving their living environment. At the same time, all workers were tested for COVID-19 in an orderly manner, including cleaning public areas, dining sections, providing standard meals, and establishing isolation shelters.

At the end of the "circuit breaker" measures, a total of 30,688 cases of Singaporean workers were infected, accounting for 95% of the total number of infections. Although the scale of the infection was relatively large, the cumulative recovery rate was as high as 61.4%, and the number of dormitory infections continued to steadily decrease. As of 1 June 2020, Singapore had conducted more than 408,000 tests.

Of course, Singapore's anti-pandemic policy, which achieved significant results in a gradual mode, was intertwined with the government's governance capabilities. In early June, the British venture capital firm Deep Knowledge Group ranked the safety levels of 200 countries and regions during the COVID-19 pandemic. Based on several indicators such as isolation effectiveness, prevention, control and detection capabilities, and government efficiency, Singapore ranked fourth globally.

Singapore's efforts from a whole-of-government approach and effective dynamic governance, as well as a cooperative society, were affirmed by the World Health Organization. It became another "anti-pandemic model" besides China's.

Japan's quick response model

On 15 January 2020, the first COVID-19 case was confirmed in Japan, and the pandemic situation continued to intensify. In response to this public health incident, the Japanese government established the "New Coronavirus Response Headquarters" on 31 January. The Prime Minister personally led the team, coordinated with the Cabinet and jointly promoted pandemic efforts with the Chief Cabinet Secretary and the Minister of Health, Labor, and Welfare. This was also the highest response level to public health emergencies for Japan.

On the morning of 28 January, the Japanese government convened a "council meeting" attended by all cabinet ministers following Article 1 Item 2 of the Cabinet Act and Infectious Diseases Act. Through a cabinet resolution (the highest administrative decision responsible to the Cabinet), the Japanese government mandated COVID-19 as a "designated infectious disease" with a "second-type infectious disease" response, stipulating that the Infectious Diseases Act shall not be revised by Parliament and can undertake the same emergency measures taken by the Middle East Respiratory Syndrome (MERS) and Severe Acute Respiratory Syndrome (SARS).

Specific measures included the issuance of hospitalization advice to patients or mandatory admission to hospitals by the government, allowing patients to seek consultation at about 400 designated hospitals with equipment and medical conditions across the country, restricting patients from working regardless of their nationality, and financing their bills at the public's expense. Doctors and medical institutions were also obligated to report once there were confirmed cases. Departments in charge of the Fire Protection, School Safety and Health Law, as well as other related departments, were to do a good job in coordinating with one another, and immediately promulgate the implementation of laws and measures of a "designated infectious disease."

In the early stage of the pandemic, the Japanese government held three National Security Conferences and Cabinet Meetings respectively. According to relevant laws, it was decided that foreigners who had stayed in Hubei and Zhejiang Provinces and held passports issued by the above two provinces, as well as passengers and crew of the international cruise ship "Westerdam," which departed from Hong Kong were barred from entering except under special circumstances. Based on the "decision" and "approval" of the above meeting, the Ministry of Justice issued an announcement "On the Application of Article 5, Paragraph 1, Item 14 of the "Law on Exit Management and Refugee Recognition", and implemented measures to restrict the above groups.

Apart from promptly setting up a command agency and issuing a response plan, the Japanese government took steps in pandemic prevention and control that will bring about the smallest possible impact on social and economic operations and reduce the spread of the pandemic without a "lockdown." The Japanese government was highly committed to discovering links between groups of infected persons as soon as possible and achieving "complete blockage" of the pandemic through effective control. In the absence of effective therapies and vaccines, the relevant measures taken by the Japanese government effectively delayed the infection peak and prevented the pandemic from spreading.

With the rapid development of pandemics overseas, Japan has further strengthened border management and quarantine since March and implemented strict quarantine and observation measures for travelers originating from key areas to Japan. The Ministry of Health, Labor, and

Welfare has specially set up a "Health Tracking Investigation Center" requesting shipping and aviation companies to cooperate and conduct border quarantine measures such as broadcasting on board and issuing health collection cards. At the same time, quarantine kiosks such as airports are required to be equipped with infrared thermometers to conduct preliminary inspections of travelers entering the country and strived to reduce the number of infected people at the source.

On the other hand, in response to the nationwide panic caused by COVID-19, central and local government leaders in Japan have issued notices to the respective populace, calling for their calm and scientific response to the pandemic. In addition, the Japanese government released relevant data on the pandemic in a timely manner to avoid social panic. Since Japan's first infected case at the end of January 2020, the Ministry of Health, Labor, and Welfare has been sharing COVID-19 data daily. It includes the patient's approximate age, place of residence and mode of transport, links between patients, and its investigation.

On 5 February, the National Centre for Infectious Diseases, with approval of the Medical Research Ethics Review Committee, decided to establish research on "Contributing to The Development of Countermeasures Against 2019-nCov Infection," and provide relevant information on issues related to research privacy and other issues on human-targeted medicine to obtain societal support and understanding.

In the face of the challenges posed by COVID-19, the policies and measures introduced by the Japanese Abe government within a limited timeframe have generally achieved relatively significant results. According to the statistics in mid-May, the number of deaths caused by the epidemic per 1 million people in the world is 258 in the United States, 584 in Spain, and only 5 in Japan. The anti-pandemic effect is the best in the Group of Seven (G7). As of 29 May 2020, the number of infected people in Japan was 16,719, while the number of deaths was 874. Compared with European and American countries, Japan has achieved remarkable results in fighting the pandemic. Therefore, this "Japanese model" has also been affirmed by the UN Secretary-General Guterres.

8.1.4 Freedom of Fighting Against the Pandemic with False Information

COVID-19 has caused a global crisis and is a test of governments' leadership. Since there are no ideal means to fight against new pathogens, countries were forced to make difficult choices on how to respond to the pandemic, which creates different anti-pandemic models. However, the leaders of the United States failed this test, and the Trump administration turned the crisis into a tragedy. Under the "freedom (model) of fighting the epidemic" in the United States, its failure to fight the epidemic is shocking.

On 21 January 2020, the first confirmed case of COVID-19 appeared in Seattle, Washington, which was an imported case. On 28 February, the US Centers for Disease Control and Prevention (CDC) announced that Northern California had its COVID-19 case of unknown origin in the United States. This patient did not leave the United States, was not in contact with other confirmed patients and was not isolated before the diagnosis. This means that COVID-19 may have entered the stage of "community transmission."

A day later, the Centers for Disease Control and Prevention of the United States reported the first COVID-19 death in the United States. It was a locally transmitted case from a nursing home near Seattle. The CDC also reminded the public that COVID has begun spreading on a large scale in American communities. On 11 March, the number of confirmed cases in the United States exceeded 1,000, which became a turning point in the development of the pandemic in the United States. On March 13, Trump finally declared that the United States had entered a state of national emergency and issued a federal level "social distancing" guide to remind the public to maintain social distancing.

However, in the face of COVID-19, Trump tried his best to play down the harmfulness of the pandemic in the early stage because of concerns that it would have an impact on economic development and undermine his reputation. According to statistics, in the 70 days after receiving China's first notification of the pandemic, Trump played down the severity of COVID-1934 times and asserted that the pandemic would "miraculously" disappear. It is precisely because of Trump's "liberal" thinking and unscientific remarks there was a delay, confusion, and failure of the US federal government's response to the epidemic. It missed the best window for pandemic prevention, and made the United States at one point a country with the largest number of confirmed cases and deaths from the virus.

In early March, the United States had entered the stage of a community outbreak, but Trump continued emphasizing that the risk of infection was very low for the general public and that COVID-19 was a deliberately fabricated "scam" by the Democratic Party and a weapon to attack him. He made speeches with political agenda in mind, rather than a medical and scientific perspective to fight the epidemic. He publicly stated that COVID-19 was no different from the ordinary flu. If you wash your hands frequently and pay attention to hygiene, even if you come into contact with an infected case, it will not cause any impact.

The downplaying of the pandemic and economic supremacy made the Trump administration very passive when managing community transmission of COVID-19. This was also reflected in the government's deficiency in virus detection and protective material procurement. According to a survey conducted by The Atlantic Monthly, as of March 8, the United States tested 3,201 people. However, during the same period, the number of tests conducted in Italy and South Korea were 49,937 and 189,236 respectively. Tabulated in proportion to the population, the United States, Italy, and South Korea tested 9 cases, 826 cases, and 3692 cases per million people respectively. South Korea and the United States reported their first confirmed cases at the same time (January 21 and 20), but in about 50 days, South Korea had 410 times the number of tests per million people in the United States.

In the face of internal criticisms on outbreaks of local transmissions, Trump deliberately referred to COVID-19 as the "Chinese virus" to divert the public's attention and intensify racial hatred. This was part of its campaign strategy aimed to transfer the anger of casualties and economic losses caused by COVID-19 to a foreign opponent that many Americans have been wary of. On the one hand, this kind of racial discrimination has impacted Asian Americans and made them targets of racial violence and discrimination. On the other hand, it has harmed other ethnic groups in the United States without an adequate, objective, and scientific understanding of COVID-19. The harm of the virus has cost many lives.

Trump's remarks on the medical aspects of COVID-19 ran contrary to scientific knowledge and were particularly unhelpful for the public and its discussions related to the understanding of COVID-19. In April, some of his right-wing supporters protested against home quarantine, demanded the economy's reopening, and even created anti-medical and scientific slogans such as "This is a fake crisis" and "Fauci was wrong." He referred to these home quarantine protesters on social media as "great people" and said he thought that they were being patriotic and responsible.

In 2020 COVID-19 reports, news, information, remarks, and commentaries, which were related to China and driven by liberalism, appeared to be in a chaotic intertwined state of anti-science and false information. As a result of the "concealment of virus theory" and "made in China theory" mooted during the early days, as well as saying such as "masks are useless," "infection rate among Americans is low," "opposing home quarantine and rejecting China's anti-pandemic model" narratives, anti-scientific liberal reports were trending, especially those published on Fox News' website and social media channels such as Twitter, as well as posts by anti-China politicians on Facebook.

On the other end of the liberal's anti-science stance is Trump's disrespect for scientific advice given by public health experts in response to COVID-19, as well as his pandemic responses that often contradicted professional opinions. Such downplaying of COVID-19 without a scientific basis sent conflicting signals and created additional obstacles for experts in their anti-pandemic efforts.

Apart from marginalizing the standing of American infectious disease experts, Trump also contradicted medical common knowledge and bypassed experts to directly recommend drugs and treatments to the public. In an unprecedented move, he strongly recommended the use of hydroxychloroquine, whose efficacy was unproven and later advised by the World Health Organization against its use for COVID-19 treatment. Trump even set an example by declaring publicly that he was taking the drug.

Trump used political manipulation and other means to undermine the tradition of political neutrality such as the independence and professional ethics of technical experts in the administration. He placed political loyalty above all, weakened the power of professionals in federal agencies, and forced scientists to serve the political will of leaders. He disbanded the Global Health Security and Biodefense unit, which resided under the National Security Council until 2018, and constantly attacked those who disagreed with his remarks or pointed out his mistakes, which increased the opportunity costs of experts expressing their professional opinions. His hostility towards professional knowledge had coerced many of the most capable and experienced federal employees to resign.

The US Center for Disease Control and Prevention, which used to be the world's leading disease prevention and control organization, was heavily impacted during the pandemic and suffered disastrous failures in its testing and policies. The National Institutes of Health played a key role in vaccine development, but it was excluded from many key government decisions. The U.S. Food and Drug Administration was also shamefully politicized. They succumbed to government pressure and did not adhere to scientific evidence. To a certain extent, Trump undermined the people's trust in science and government, and the effects would last longer than the pandemic. The government no longer relied on professional knowledge but resorted to ignorant "opinion leaders" and fake experts, who covered up the truth and encouraged the spread of outright lies.

From the history of human pandemics, there has not been any major infectious disease unresolved by medical advances and scientific methods. Infectious diseases can never be solved by slogans. Instead, they need to rely on professional medical knowledge and scientific control to effectively manage them.

As of 8 October 2020, regarding data from Johns Hopkins Centre for Systems Science and Engineering, the number of patients and deaths from COVID-19 in the United States ranked first in the world, far exceeding China and other countries with larger populations. The fatality rate in the United States was more than twice Canada's and nearly 50 times higher than Japan, which has an aging population and therefore has a larger proportion of people at a high risk of contracting Covid-19. It was also nearly 2,000 times higher than that in low- and middle-income countries such as Vietnam. The Trump administration has turned the crisis into a tragedy. Under the United States' "freedom to fight the epidemic," its failure to fight the pandemic is shocking.

8.2 Three Problems Behind the United States' Inability to Fight the Pandemic

In March, the COVID-19 pandemic broke out in the United States.

By the beginning of October, more than 760,000 cases were confirmed in the United States, with total deaths exceeding 210,000. Although the growth rate has declined, the overall situation remains unstable. Apart from New York and other places, where the pandemic was relatively severe in the early days, and life is gradually returning to normal, other places are still waiting for an improvement in the situation.

Materially, the United States is the top global power. No other country can compare with it in terms of its political, economic, military, technological strength, and dollar hegemony. In terms of its population, the population of the US is about 330 million, which is less than a quarter of the population of China. It should be easier and more efficient for it to fight the pandemic, compared to China.

However, the reverse is true. While the pandemic in China has entered a normalized stage in its prevention and control, the United States has entered a prolonged phase. As the world's most powerful country, why is the United States so weak in fighting the pandemic?

The economically powerful US anti-pandemic issue is no longer purely material, but it is a political, social, and cultural issue.

8.2.1 When the Pandemic Overlaps with the Election

The United States is a multi-party country. The obvious advantage of a multi-party system is that the people can choose who will govern. When a certain political party is in power, the opposition party can supervise it to make sure it does not deviate too much from the social contract recognized by everyone. If the ruling party goes too far, it will be not re-elected. One of the unavoidable drawbacks of the multi-party system is the unscrupulous means used by the two parties to fight for power, including malicious slander and deceitful propaganda.

The year 2020 coincided with the U.S. election year. The large-scale COVID-19 outbreak in the United States, which went out of control within a short time, had undoubtedly become the biggest black swan event in the U.S. election year. At the same time, the conflicts in politics, power, discourse, and ideology have led to entanglement in political campaigns and anti-pandemic battles, severe disruption to the prevention and control of the epidemic, making it impossible for the government to take effective and timely measures.

For example, the pandemic briefing, which is meant to be an important window for the United States to prevent and control the pandemic, should objectively convey to the public the facts of the pandemic, to mobilize the Americans to work together to defeat it. In reality, it was designed by the White House Information Office to become Trump's campaign stage, it included meeting small entrepreneurs, recovered COVID-19 patients, volunteers, and other groups of people, and letting them come on stage to express their gratitude to Trump.

In addition to turning the pandemic conferences into a campaign platform, Trump asked the states to resume work and schools as soon as possible while the pandemic was still spreading and before effective protective measures were introduced. This was done to prevent the economy from holding back the elections. These decisions, based on election considerations, resulted in the ineffective control of the pandemic.

This also resulted in Republican governors following suit. Georgia Governor Brian Kemp even took the Democratic Atlanta Mayor Keisha Lance Bottoms, who asked everyone to wear masks, to court in mid-July, and said that the restrictions imposed on the resumption of work violated the governor's order (they later reached a consensus on this matter, and the governor withdrew the application in August).

In addition, Georgia's North Paulding High School, which already had students and staff who were confirmed to have COVID-19, required students to attend classes without wearing masks. Those who did not comply remained on campus for observation or faced expulsion. Any teacher and student who criticized the school on media faced disciplinary action.

The situation in the universities is similarly not optimistic. According to the New York Times statistics, as of September 10, there were at least 88,000 cases at the 1,600 four-year public and private universities which were investigated. Among them, 1,889 cases of infection occurred from mid-August to September at the University of Alabama in Tuscaloosa. There were more than 1,200 cases at Iowa State University.

In addition, mutual attacks between the Democratic Party and the Republican Party on the pandemic issue led to the fragmentation of efforts.

Trump attacked the Democratic Party's incompetence and chaos, because the governors of the most severely hit states in the United States, such as the hardest-hit areas in New York, California, and New Jersey, were Democrats. He also accused the Democratic Party of trying to politicize the virus and attempting to use the seriousness of the pandemic as a political tool to attack him.

In Trump's view, it was because of the Democratic Party's loose immigration policy that a large number of outsiders have brought in the virus. As a result, Republicans represented by Trump attacked Democrats for disregarding the safety of the American people and leaving more American people out of work and famine.

On the other hand, Democratic officials criticized Trump for habitually lying, committing mistakes while fighting the pandemic, attempting nothing and accomplishing nothing, conducting a show election, and making arbitrary decisions. They also attacked Trump for disregarding civilian lives and restarting the US economy. In turn, they advocated putting the people's health above all, banning incoming travelers, and suspending the economy to fight the pandemic.

With the election and pandemic overlapping, the partisan infighting in the United States has eventually resulted in the pandemic situation deteriorating again.

8.2.2 Cognitive Distortions under Anti-Intellectualism

Regarding American anti-intellectualism, perhaps the most famous assertation comes from Richard Hofstadter's masterpiece *Anti-Intellectualism in American Life*. It said "Americans hate and doubt the intellectual life, and those who represent it."

The book also pointed out a romantic belief and mentality of Americans, which is that popular democracy should support that "innate, instinctive and folk wisdom are in a superior position, overriding the literati and wealthy's education, learning and profiteering from knowledge." Practical experience is more important than imaginative thinking, and instinctive emotions defeated the rationality of anemia.

"Just as evangelicals refuse to accept learned religions and formal pastoral systems and prefer inner wisdom and direct listening to the voice of God, supporters of equality politics reject well-trained leaders and prefer the inner feelings of ordinary folks to directly understand the truth. This preference for ordinary people's wisdom grew in the most extreme statements of democratic doctrine and turned into offensive anti-intellectualism."

Hofstadter believed that the American anti-intellectual tradition, which is the dislike and questioning of knowledge, reason, and intellectuals and elites who represent these things, as well as the greater emphasis on common sense in life, began earlier than the existence of the

American national identity. Hofstadter cited many contemporary examples, including attacks on universities by right-wing forces.

The problem of anti-intellectualism exists everywhere, and Trump's anti-intellectualism was manifested incisively and vividly during the pandemic. After the outbreak in the United States, Trump first publicly stated that COVID-19 is just a cold, and the virus will soon disappear.

In April, Trump even created an unprecedented disinfectant treatment, suggesting to the medical community a study on injecting disinfectant into the human body to eliminate the new coronavirus or use ultraviolet rays to irradiate the patient's body to achieve a therapeutic effect. He was trending on the list of Weibo hot searches at one point. Soon thereafter, he tried his best to recommend hydroxychloroquine, which had unproven efficacy, and he even used the drug publicly. He was later ordered to stop consuming it by the World Health Organization.

In August, Trump appointed neurologists, conservative think tank experts, and Atlas, who is not an infectious disease expert, to join the White House's pandemic prevention team as the latter is in public policy (for instance, "herd immunity" which has been abandoned globally) and is consistent with Trump's view on economic issues. However, the real infectious disease experts Burks and Fauci were marginalized. As the most trusted epidemiologist in the United States, Fauci was restricted by the government in popularizing pandemic control and prevention on mainstream TV stations, so he could only speak on social media.

Trump forced his government agencies to follow his direction. Among them, the Federal Food and Drug Administration proposed to use the plasma of COVID-19 patients as a treatment method and stated that its efficacy rate reached 35%. However, experts pointed out that the data was without basis. In addition, to reduce the number of COVID-19 cases, Trump instructed the Centers for Disease Control and Prevention to propose a guiding principle, which was that no testing was required for asymptomatic cases. Although the total number of cases decreased in this instance, the actual cases did not (asymptomatic infections were not included in China's early calculation of cases). According to Trump's logic, there is no case without testing, and no virus means that there is no case. Such ignorance means there is no need for fear.

Famous contemporary western social theorist Goran Thurborn divided ideology into three levels, cognitive ideology, normative ideology, and transformative ideology, in his book titled *The Ideology of Power and the Power of Ideology*.

Among them, cognition and ideology involve human cognition of the world. For instance, questions such as what exists and what is non-existent, and what is factual and what is not. Trump's anti-intellectualism is precisely a distortion of cognition and ideology. He used "playing with facts" as an ideological tool to directly challenge the facts, and even distorted the facts to challenge the truth, thereby preventing the truth from being revealed.

As a result, Trump's belief, and desire that "COVID-19 will not cause disaster to the United States" formed at the beginning of the pandemic influenced his behavior during the entire fight against the epidemic and distorted his choice of scientific facts. It was also tied to his cognitive preference. His selective cognitive preference for scientific facts has strengthened his initial understanding of "viruses will not cause disasters" to form a distorted "belief and desire," which was integrated into a distorted closed loop of cognitive ideology.

What is absurd is that this "belief and desire" has also received the automatic cooperation of a considerable number of Americans, and both sides have reached a "cognitive coordination" mechanism. As a result, absurd remarks such as "COVID-19 is a common flu and a well-planned scam," "hydroxychloroquine is a COVID-19 drug that can change the rules of the game" and "injection of disinfectants can kill COVID-19," have surfaced. They may appear to be justified and natural for the people in this show, but to outsiders, this is unbelievable.

This type of anti-intellectualism, which dislikes and resists scientific authority, as well as uses and betrays science, has undoubtedly had a huge impact on the progress, prevention, and control of the pandemic.

8.2.3 Abuse of Freedom from the Risk of Disorder

Fighting the pandemic is a nationwide activity. In the United States, apart from the worsening situation caused by the elections, part of the resistance to the United States' responsibility stems from the inefficiency of its democratic governance system.

Democracy and freedom are known as the spiritual pillars of the capitalist state government. In reality, their relationship under capitalist conditions is not coordinated. Freedom cares about individual rights, while democracy cares about the control of individual and collective behavior. This means that there is a contradiction between them, and they are in a state of tension under certain conditions. The enthusiasm and obsession with freedom have subtly affected Western societies, leading to the disorderly operation of democracy.

This disorder is prominent in American politics. For example, the executive branch defies the parliament and often uses the "state of emergency" as an excuse to evade established rules and procedures. The abuse of freedom is also common among ordinary Americans, which is reflected in the "Prevention of Epidemics at Home" parade.

At the same time, this pursuit of freedom is also reflected in the pursuit of privacy protection of the American people in the prevention and control of the epidemic. Take case tracing as an example. Case tracing refers to contacting those who might be infected and those who were in contact with them, and placing them under quarantine for 14 days to prevent them from infecting others. In mainland China, community tracing is often carried out by the community. In Taiwan, it is carried out by the chief (the lowest-level government agency) and his subordinates. However, since the United States does not have this level of government agencies, the city or county can only hire a dedicated person to call in and track it.

However, tracking has been highly unsuccessful in the United States. The reason, of course, is related to insufficient detection methods and prolonged tracking time, and the population's unwillingness to cooperate. For example, China uses a health code (including various ID card information and facial recognition functions) to track everyone. It also relies on bracelets and GPS on mobile phones to track patients like other Asian countries and regions. This is unthinkable in the United States. Therefore, American telephone commissioners always rely on some methods, for instance, by first asking whether there is enough food, whether there are unemployment benefits and whether baby diapers are needed to first gain the trust of the other party.

Nevertheless, case tracking is still very difficult. According to a New York Times report, a follow-up survey in Los Angeles in a week in July found that more than one-third of patients did not answer the phone, and half of them refused to disclose who they had contact with.

Therefore, the disorder risk caused by the proliferation of freedom threatens the efficiency of governance. Order is a prerequisite for efficiency and provides a standardized guarantee for greater efficiency. Without the assurance of order and rationality, the government's prevention and control measures will not be efficient.

On the other hand, social cohesion reflects a society's ability to coordinate internal conflicts and mobilize members of the society. In the face of major crises and incidents, this kind of social force that eases conflicts and promotes cohesiveness is very important. However, American democracy, based on the value of individualism, is ineffective in promoting social cohesion.

Objectively speaking, the development of American democracy is benefiting from such individualism that pursues autonomy, but it has also gradually plunged society into a quagmire of self-centeredness, and the gap between public and personal life has gradually widened. The United States tries to overcome its individualistic inclinations and establish associations to shift and harmonize the stands between public life and personal life. However, there is a conflict in values between individualism's emphasis on the self and society's emphasis on the collective. Unless there is a strong social integration system, its integration effect on individuals will be greatly restricted.

However, the conditions of capitalism are not equipped for this type of integration system. Capitalism has strong liquidity, and social relations formed under its influence are in an unstable state, and this state affects the organizational effectiveness. Individuals cannot bear the burden brought about by the unstable structure of society, and their unreasonable demands will further impact the foundation of social solidarity. Therefore, it is difficult for the population to gather strength to fight the pandemic, and even the state governments' handling of the pandemic cannot escape the barriers of individualism. In comparison with the mutual assistance and cooperation of various provinces, autonomous regions, and municipalities in China, the US state governments are more like competitors competing for medical supplies.

In short, the US government's sense of weakness in response to the pandemic prevention and control is not unfounded. Party politics, anti-intellectualism, and free abuse of freedom are important reasons that hinder the prevention and control of the pandemic.

How to find balance or get out of the predicament is a question that the United States must think about. Otherwise, even if there is no COVID-19 outbreak, the next disaster will come. By then, the United States and its people may have to pay a heavier price.

8.3 The Collapsing American Healthcare

The US pandemic prevention and control has attracted global attention during the global COVID-19 pandemic. What is the impact of the high fatality rate on the US medical system?

One argument is that, compared with China's "guarantee of life but not your fortune," the United States firmly implemented the reverse, directly overriding the "collapse of the healthcare system" and realizing the conclusion of "a large number of people dying without treatment due to gaps in the system." As a result, the American people concluded that "the American medical system will not collapse." Undoubtedly, this conclusion sounds absurd, but it may also further reflect the public's helplessness and ridicule of the American medical system.

8.3.1 American Medical Care the Chinese Cannot Comprehend

Adhering to the centuries-old tradition of liberalism, individualism, community autonomy and elite guilds, medical care in the United States does not operate in a comprehensible way for the Chinese people.

As early as 2014, the total health expenditure in the United States was as high as 3 trillion US dollars, accounting for 18% of GDP. At the same time, there were still about 15% of the population not covered by any medical insurance, and their average life expectancy was the second-lowest among OECD countries. In 2018, the total health expenditure in the United States still accounted for 16.2% of GDP, ranking first in the world. In contrast, China's total health expenditure only accounted for 4.6% of GDP, ranking 145th.

If the US medical industry was regarded as an independent economy, it will even become the fifth largest economy in the world, surpassing France, the United Kingdom, Russia, and Brazil. To understand the bizarre phenomenon of the U.S. medical industry's share of GDP, we must first understand the US medical system.

In the medical systems of China and many countries globally, doctors are subordinate to hospitals as organizations. The hospital is responsible for the entire process of outpatient service, examination, pharmacy, charging, surgery and hospitalization, and most doctors are based there. In the United States, doctors are the center of the medical system, and the main ones, comprising 630,000 doctors, are all clinic doctors.

Community clinics are the cornerstone of the American medical system. They not only provide primary medical services but also solve the most common and frequently-occurring diseases. Unlike almost all surgical operations in China, which are performed in secondary and tertiary hospitals, 68% of surgical operations in the United States are performed in clinics (including day surgery centers). Patients with more urgent or more serious illnesses are generally diagnosed and treated in outpatient centers (multi-department clinics), emergency centers, while more serious cases are treated in acute care hospitals (Acute Care Hospitals). However, the service provided by this kind of hospital is mainly hospitalization, with a relatively short hospitalization time of an average of 5 to 6 days. Upon a patient's discharge, he or she may also be transferred to a rehabilitation hospital and a long-term care hospital. There is no such difference in hospitals in China. Most hospitals combine these three types of services.

Doctors in the United States have full autonomy to choose the place of practice. The result of self-selection is that 84% of community doctors only practice in one location, but there are also 25% of clinics, amounting to about 270,000, which are joint clinics of specialist doctors. There is only one doctor in 51% of clinics, and the number of doctors in this type of clinic accounts for 18% of the total number of doctors. Thirty-eight percent of clinics have 2 to 5 doctors. In other words, nearly 90% of community doctors practice in clinics that have no more than 20 people.

According to the agreement with the insurance company, the patient sees a doctor at a fixed clinic over a long period. Doctors are generally general practitioners, but there are also specialists. They have long-term management of the health records of patients in this community and even the patient's family for several generations.

In addition, the United States implements a system of complete separation. The clinics and pharmacies are completely independent. Many more complicated inspection departments also operate independently, and medical services are quite fragmented. All clinics have to establish a cooperative relationship with one or two hospitals. If the general practitioner cannot solve the problem, they must refer it to a related specialist. Specialist doctors who do not have the conditions for complex surgery will continue to introduce referrals to the contracted hospital or borrow the hospital resources to perform the surgery themselves.

In addition to the doctors who operate clinics, only a small number of doctors are hospital employees. The remaining doctors are still independent free practitioners (about half of the independent doctors will also form a "doctor group" to practice as a group). For freelance doctors, the hospital is just their partner. To them, the hospital is like a "shared office" place. Doctors simply "borrow" the hospital's surgical equipment, instrument beds, and nursing staff to diagnose and treat patients. The diagnosis and treatment fees are collected by the doctor separately, while the other service fees are collected by the hospital.

Therefore, many general hospitals only hire some young doctors in the emergency room, and specialized departments are only equipped with nurses and medical equipment for the contracted attending doctors to "borrow." Regardless of whether it is a general practitioner, a specialist, hospital, pharmacy, or examination center, all patient appointments, referrals, dispensing, operations, and checkouts must be placed under the insurance provisions, restrictions, and contact, otherwise they will not be paid.

Therefore, due to high costs and restrictions on terms, the average American sees a doctor only 4 times a year, while the median number of times a person sees a doctor in other OECD countries, comprising 36 more developed countries, is 6.5 times a year. Japan is 12.9 times.

8.3.2 Unique Medical Insurance

For advanced economies, medical insurance in the United States is a special case. Unlike European countries, Canada, Japan, and other developed economies, which generally establish social medical insurance systems covering the entire population, the US government-led social medical insurance focuses on protecting the elderly (Medicare) and the disadvantaged (Medicaid), while the medical insurance for the working population is covered by commercial insurance agencies.

Apart from government medical insurance, commercial medical insurance in the United States is extremely fragmented. However, it also constitutes a game among individuals, employers, insurers, and medical providers (doctors, hospitals, and pharmacies).

It is worth mentioning that there are historical reasons why American medical insurance is mainly commercial insurance. American politics has three distinctive features, namely federalism (decentralization), limited government, and liberal ideology. Social services, including medical services, have always been under the jurisdiction of the state government rather than the federal government. Some employers provide basic medical insurance for their employees, which is more in line with American political and economic concepts. The government has been intervening in medical insurance and had not received sufficient support for a long time, making it difficult to legislate at the federal level.

During World War II, the United States implemented wage control. Employers could not raise wages, so they turned to provide medical insurance and other benefits to attract talents. In 1954, the United States imposed tax exemptions on the cost of medical insurance provided by companies to their employees. Since then, employers have become the main channel for American nationals to obtain medical insurance.

When people generally hear about the extremely high or low medical charges in the United States, it often depends on what kind of insurance one has – good insurance, bad insurance, or being uninsured.

If it is a government employee, a large commercial company, especially a large high-tech company in Silicon Valley, it will have strong negotiation and high paying ability. At the same time, it is good for employees. Employees can enjoy excellent insurance coverage, such as monthly individual payments of 400 US dollars, company contributions of at least USD 1,200, with the insurance clause covering a wide range. For small employers or even self-employed workers, there is no advantage in the negotiation, so the insurance purchased is often not effectively realized when it is needed. As a result, the out-of-pocket ratio is very high, and many medical treatments preclude many situations.

But, in any case, according to public data, middle-aged workers in the United States have to spend US$400–500 per month on insurance (US$5000–6000 per year) to be considered as having insurance coverage. At the same time, 11% of the population (about 35 million people) cannot afford commercial health insurance (16% before Obamacare). For the lowest class, you can only pray that you do not get sick. If you are sick, you can only go to the emergency center with the government's support, but it is only interested in saving lives, not treating chronic diseases.

One of the major characteristics of American medicine and medical insurance is the vast difference between the asking and given prices. For a simple emergency with a prescription of a few painkillers, and the clinic doctor can prescribe a "basic bill" of US$4,000. However, the insurance company will ignore it, and will only pay US$1,000 as agreed in the contract, and the patient will pay US$100 for the remaining amount.

For the uninsured, the original bill is already at a sky-high price. In the United States, personal bankruptcies caused by huge amounts of medical care account for 62% of bankruptcies.

From employers, medical providers, to insurance companies, they have to spend time and energy arguing and negotiating both before and after treatment. Hence, the third characteristic of American medical care has emerged, which is, the price is not transparent. Ordinary people are not eligible to compare prices if they do not make public announcements beforehand and send bills afterward. In addition, the flat nature of insurance means that insurance payment, out-of-pocket costs, and even the management costs and profits of insurers will ultimately be borne by the people.

The combined efforts of all parties have finally created the current medical system in the United States. The per capita annual medical expenditure exceeds US$10,000, which is 2.6 times the average of US$4,000 in OECD countries. However, the average life expectancy is the second-lowest among OECD countries.

8.3.3 How did the Pandemic Disrupt American Healthcare?

The latest report of the American Hospital Association shows that in 2018, all community hospitals in the United States had about 800,000 beds, comprising less than 100,000 ICU beds and less than 70,000 adult ICU beds. Currently, the United States has 62,000 full-function ventilators and 98,000 basic ventilators, and a national strategic reserve of 8,900 units. In addition, in 2018, there were 76,000 full-time respiratory therapists and 512,000 intensive care nurses in community hospitals across the United States.

In May 2020, the total number of confirmed cases in the United States has exceeded 1.66 million. Assuming that the proportion of severe cases is 15%, approximately 249,000 ICU beds are needed. At present, the upper limit of adult ICU beds in the United States is 70,000. Even if the children's ICU beds in the United States are also used for COVID-19, it is far from enough for the United States to treat it. According to the experiences of other countries, at least 10% of hospitalized patients need to use a ventilator. Assuming that the hospitalization rate is 10%, 166,000 units are required. California law stipulates that a respiratory therapist can treat up to 4 patients at the same time. Assuming that the United States follows this standard, 415,000 respiratory therapists are needed, which is far more than the actual number of full-time respiratory therapists in the United States.

Any medical resources required for COVID-19 could no longer meet the actual needs of the United States at the time. As such, who can say that the US medical system has not already collapsed?

In addition, the unique medical insurance system in the United States has led to a large number of uninsured people, which has exacerbated the challenge of dealing with the coronavirus. The survey shows that at least 27 million Americans do not have health insurance. Also, as millions of people lose their jobs, this number will rise. Without insurance, patients may not be able to pay high fees due to emergency room visits, forcing the hospital to bear bad debts.

Although the United States spent trillions of dollars on health care, most of it was wasted. A study conducted in the *Journal of the American Medical Association* in 2019 found that at least $760 billion was wasted on unnecessary medical expenses, which was higher than the US disbursement on primary and secondary education. The US needs to spend more. These funds were consumed by bureaucrats who have to code and bill every action taken by doctors, and doctors and hospital administrators paid far more than their European counterparts.

It can be said that the US medical system flaws have been exposed due to COVID-19. For decades, the United States has been trying to figure out how to change its rigid medical system. Can this pandemic trigger a campaign that will ultimately make it fairer? It is possible that COVID-19, just like the suffering caused by the Second World War to the British national medical system, can also change American hospitals forever. Of course, it is also possible that the system will face further roadblocks.

8.4 Inspirations from the British Anti-Pandemic Efforts

When the First World War began, the British public was optimistic that the war would "end before Christmas." Facts later proved that this was just wishful thinking. The First World War lasted four years and eventually ended when the participating countries were exhausted. History is always surprisingly similar. In the face of the global COVID-19 outbreak which broke out in early 2020, the British public also agreed that the COVID-19 in the UK will end soon, but just like then, the facts once again proved that it was just the public's wishful thinking.

On 31 January 2020, a confirmed case appeared in the UK for the first time. Subsequently, the British anti-epidemic experience went from herd immunity to a public outcry, with Prince Charles and Prime Minister testing positive, resulting in a tightening of the British pandemic prevention policy. By the end of 2020, the UK will face the pressure of a second pandemic peak. However, just as the vaccine gave hope to eradicate the pandemic, British officials issued an emergency warning that a new highly contagious variant of the new coronavirus began spreading in the UK. At the same time, many countries have announced the suspension of flights to the UK, which once again aggravated people's concerns and added "uncertainty" to the already challenging fight against the pandemic.

Of course, as COVID-19 is a new disease, uncertainties are real, and some mistakes are inevitable. Also, there are differing views among epidemiologists. It is not just the UK that is facing problems. Scientists often disagree, as do clinicians and public health experts. However, the development of the pandemic in the UK lags behind most European countries. Despite lessons learned from Italy and elsewhere, Britain's mortality rate is among the highest in countries outside the United States. Management errors in procurement and distribution exacerbated political errors, leaving the UK lacking key resources such as ventilators, testing capabilities, and personal protective equipment.

It can be said that the British fight against the pandemic is another visible failure outside the United States. The United Kingdom's anti-pandemic experience has been full of ups and downs and challenges. However, when reviewing the events that occurred at every point during the epidemic, from the introduction and entry of herd immunity to the pressure of the second pandemic wave peak, as well as the arrival of new coronavirus, history has always been inspiring in an upward spiral manner.

8.4.1 The Opportunity that Cannot be Missed

One of the biggest features of the UK's anti-pandemic is that it is "half a beat" slower than other countries. Although the isolation and blockade strategies adopted by European and American countries were later adopted by the British government, the problem was that the adoption of these measures was not efficient and forward-looking and seemed to be "forced out" by civil society and the seriousness of the pandemic.

The British government's initial strategy for responding to the epidemic is "containment-delay-mitigation-research." In early March, when the spread of COVID-19 accelerated in the UK, the government announced that the anti-pandemic had entered a postponement stage from containment, but there were few new actions on their part. This has delayed the opportunity to contain the pandemic to a certain extent.

Among them, at the beginning of March 2020, the British National Health Service (NHS) notified British residents who had overseas travel history or flu symptoms in the past 2 weeks or had contact with confirmed COVID-19 cases to call 111 and requested them to isolate home and not go directly to the clinic or hospital. Before that, many British people who had returned from European travel had already developed COVID-19 symptoms, and Britain has entered an outbreak.

On 12 March, the British government made a disastrous decision to stop testing and tracking the coronavirus. However, government officials insisted that this was to prepare the National Health Insurance system for disasters. On 14 March, 1,140 cases were diagnosed in the UK. The British government publicly admitted that the actual number of infections in the UK may be between 5,000 and 10,000. The NHS (British National Medical Service System) announced that it will no longer test people with mild symptoms. In comparison, on 23 March, Germany's virus detection capacity has reached 103,000 times a day, while the UK's detection capacity has just increased from 5,000 times a day to 10,000 times a day.

In addition, although the school's overseas trips were canceled, elderly people who were already sick were not recommended to go on cruises. However, the "hard" anti-pandemic measures such as the cancellation of large-scale public events and closure of campuses that many expected and called for were not included because "the timing was not right."

Prime Minister Boris Johnson emphasized that it was very important to grasp the timing of implementing pandemic prevention measures. Taking reference from the opinions of scientific and medical advisers, Johnson said that the UK does not need to close schools or cancel large-scale public events, such as sports events. These were the decisions made after weighing many factors.

The British experts assessed that if these measures were taken, the cons would outweigh the pros. The chief medical adviser of the government said that the fight against the pandemic was a protracted battle. Taking such measures prematurely can easily cause mental and physical fatigue, which is not conducive in the fight against the pandemic and may not be able to combat the vital points effectively. On the contrary, it will bring out an additional burden. However, this ran contrary to the general understanding of academia, which believed that once the pandemic enters the postponement stage, it is necessary to start adopting "social distancing" measures, including canceling large-scale events, closing campuses, encouraging, and recommending employers to allow employees to work remotely from home.

However, the measures taken by the United Kingdom during the delay phase did not play a positive role. It failed to grasp the ideal period to contain the pandemic, and the result could only be similarly disastrous. Judging from the trend of the UK pandemic, the UK will soon usher in an upswing period. In March 2020 alone, the number of confirmed COVID-19 cases in the UK surged from 12 to 25,150, with a total of 1829 deaths and a total of 135 cured, which can be labeled a disaster.

In fact, during the COVID-19 pandemic, people often compared it with past pandemics. In 1957, the Asian flu caused more than 2 million deaths worldwide. At least 14,000 people died directly from the virus in the UK, more than 9 million confirmed cases, and 5.5 million people were treated. After that, the British economy fell into recession. It was worth mentioning that in 1957, the World Influenza Research Centre (WIRC) in London formed a network for research, tracking, and monitoring of viruses with laboratories around the world.

In the early summer of 1957, the Asian flu began spreading in Britain. The first case of infection appeared in late June. The pandemic broke out in August and spread to all parts of the country within a few weeks. There was no escape. The pandemic peaked in mid-October. There were limited repetitions during winter that year. The main infected population of the Asian influenza virus was 5–39 years old, and nearly half of them were minors aged 5–14. At that time, the United Kingdom did not have a generally uniform treatment plan for the Asian flu, and community clinics were on their own. Antibiotics were mainly used, but later it was discovered that there was no benefit to them as they were used in an untargeted manner.

The Public Health Laboratory Service of the United Kingdom (PHLS) is responsible for checking against cases of various infectious diseases, and notifying hospitals, clinics, and doctors of pandemics and pandemics conditions thereafter. The Epidemiological Observation Department of the Royal College of General Practitioners (RCGP) has similar functions. During the fall of 1957, after the Asian flu entered the UK for some time, PHLS chief J Corbett McDonald wrote to Ian Watson, chief of the Epidemiological Observation Division of RCGP, complaining that neither of the two companies had conducted large-scale studies during the outbreak. Therefore, the subsequent in-depth research on this area was relatively limited.

MacDonald sighed that there were more than 30 years to prepare for the next pandemic after the Spanish flu, the fact proved that everyone remained passive and did not plan, which resulted in the hasty response when the pandemic struck. Hence, we can only hope that the front-line medical staff can seize the opportunity and have enough information to fully explain the outbreak eventually.

Compared with Asian flu more than half a century ago, modern medical equipment and epidemiological research had made a qualitative leap. However, more often than not, we still fail to grasp the opportunity in the first instance, leading to a passive containment of the pandemic. Whether it is an individual or a society, in the face of urgent public health events, seizing opportunities, taking precautions, and planning will play an important role in the outcome.

8.4.2 Failed Herd Immunity

On 12 March 12020, Johnson officially announced that the British war against the epidemic had entered the second phase or the delay phase, at a COVID-19 press conference. Johnson said that following scientific advice, the UK will adopt "herd immunity" measures to enable 60% of the British to be infected with the virus.

Simply put, herd immunity is when enough people are immune to the pathogen that causes the disease, so that other individuals who are not immune can be protected from infection. The theory of herd immunity indicates that when a large number of individuals in a group are immune to a certain infectious disease or if the susceptible individuals are few, the infection chain of those infectious diseases spreading between individuals will be interrupted.

The calculation of herd immunity is completely dependent on the estimation of R0. However, as the estimation of R0 is affected by different models and social environments, the calculated value of R0 will also vary greatly. In the journal *Emerging Infectious Diseases* published by the US Department of Disease Control on 7 April, the R0 of new coronary pneumonia was updated to 5.7 (95% CI 3.8–8.9), which meant that every COVID-19 patient can infect 5.7 people. This meant that immunization must be achieved through vaccination or pre-infection of more than 82% of the population to achieve herd immunity and stop transmission.

Therefore, Johnson's "herd immunity" measures, aimed at infecting 60% of Britons with the virus and suggesting that Britons bid farewell to their families in advance, attracted criticism from professionals in the medical journal *The Lancet*. More than 600 scholars from many universities and scientific research institutions in the United Kingdom issued three open letters in a row, requesting the British government changes its passive anti-pandemic measures, stating that this would put tremendous pressure on the national medical service system and bring unnecessary risks to people's lives. The error of its decision was reflected by the quick, cumulative number of confirmed cases.

Since herd immunity is intended to reduce the impact and pressure on the health system by delaying the pandemic peak to stagger the flu season, no strong prevention and control measures were taken. As a result, the Prime Minister and the Princes were infected with COVID-19, and the number of confirmed cases skyrocketed. As of 12 June 2020, the cumulative number of confirmed COVID-19 cases in the UK exceeded 290,000, ranking fourth in the world, with a confirmed case fatality rate of 14.2%, making it the country with the largest number of deaths due to COVID-19 in Europe.

In addition, a new study in The Lancet used scientific data to prove the undesirability of herd immunity once again. Specifically, researchers from Spain's Carlos III Health Institute conducted a national epidemiological study on Spain. From 27 April to 11 May 1, researchers conducted a questionnaire survey and serological analysis for COVID-19 antibodies for 61,075 people from 35,883 families across Spain and found that only about 5% of people in Spain had the antibodies. In areas with severe pandemics, such as Madrid, more than 10% of people carry antibodies. Even so, this ratio is still far below the 55% or 82% required for "herd immunity."

Based on these current findings, a review article published by The Lancet at the same time stated that the realization of herd immunity through uncontrolled natural infection was not only unethical but also impossible to achieve. Once the control measures are canceled, the second wave of the pandemic is likely to happen quickly.

8.4.3 Run-on Medical Resources During the Pandemic

In a pandemic, there was a run-on medical resources when confirmed cases continued rising at a high speed. The UK's National Medical Service System (NHS) ability to withstand the pandemic also attracted a lot of attention. While it is a public medical system, the "high mortality rate" of COVID-19 was a great contrast to the national healthcare system that the UK is proud of. Faced with a rapidly increasing number of cases, the British national medical service system was saddled in a situation of not having enough beds, especially in the intensive care unit (ICU).

At the beginning of March, according to the Daily Mail, due to the shortage of beds and equipment in the British National Medical Service System, the hospital had only 35 intensive care beds, and the elderly and weak patients would not receive intensive care. At the same time, approximately 410,000 elderly people lived in nursing homes in the UK. It is estimated that since the COVID-19 outbreak, there had been at least 12,000 excess deaths in nursing homes in the UK, making it the second-largest source of death. In early February, the British government's guidelines for the pandemic in nursing homes were "very unlikely" to be infected with COVID-19, resulting in nursing homes facing shortages of protective equipment and medicines.

On 6 April, 26 deaths and 126 suspected infections occurred in nursing homes in Enfield, London. However, nursing homes in this area could only conduct 10 tests a day, and the staff could not accept the tests until the 15th. The British government promised that all residents and staff of nursing homes in England will be tested for the virus from 29 April.

However, a nursing home in Enfield District received only 10 kits, and 8 of them tested positive. The NHS's decision to transfer some hospital patients to nursing homes was one of the causes of infections in nursing homes. In mid-March, the NHS announced that 15,000 patients would be transferred from hospitals to community health care, and some were transferred to nursing homes. A guide issued by the Ministry of Health in early April even stated that before the elderly are transferred to or admitted to nursing homes, there was "no need for a negative COVID-19 test." This attitude of "almost abandoning nursing homes," combined with the embarrassment of running on medical resources, accelerated the deterioration of the pandemic's situation.

The Western medical system has always been based on the concept of patient-centered medical care. However, in the face of pandemics, its deficiencies were revealed. Specifically, when the hospital was quickly filled with infected patients, it became the main carrier of COVID-10 and promoted its spread to uninfected people. The transportation of patients was handled by the local response team. When ambulances and emergency workers quickly became virus carriers, it also caused the pandemic to spread. Among the health workers, were asymptomatic carriers or unsupervised patients, including young people who could have died. By then, the pressure on frontline response personnel increased.

Patient-centric medical care during the pandemic is insufficient, and we need to change our thinking. In a *New England Journal of Medicine (NEJM)* article, first-line doctors put forth a reflection on pandemic prevention and control, for instance, community-centered medical protection measures should be chosen, and the solution to COVID-19 must be for everyone, not just for hospitals. Otherwise, when the hospital is far overloaded, it would become uncontrollable.

Avoiding this disaster can only be achieved by large-scale deployment of medical outreach services. The pandemic solution must be for everyone and not just for hospitals. Also, home care and mobile clinics can avoid unnecessary activities and reduce the pressure exerted on hospitals.

For example, early oxygen therapy, pulse oximeter, and nutritional products can be delivered to the homes of mild and convalescent patients, and a wide-ranging medical and health system can be established under adequate isolation conditions, and new telemedicine tools can be used for treatment. This method would control the inpatients of a certain degree of severity in its target group, thereby reducing the chance of infection, protecting patients and medical staff, and minimizing the consumption of protective equipment.

The outbreak was not only a public health challenge but also a challenge to cross-border and cross-disciplinary collaboration. It required sociologists, epidemiologists, logistics, psychologists, and social workers to collaborate. Although its fatality was not too high, it was extremely contagious. The higher the degree of medicalization and centralization of a society, the wider the spread of the virus may be.

The plague is a plague because it generally disregards the established reality of human beings. Just as all kinds of "democracies" are no longer political concepts in the pandemic but they are transformed into "states of emergency" across different countries, the virus is forcing people to re-examine their lives. Overall, the British government missed the earliest and most effective containment phase. In addition, it did not pay attention to the protective equipment of medical staff, failed to improve the detection capabilities as soon as possible, and required those with minor illnesses to isolate themselves at home (the shelter hospitals built were mostly left unused), resulting in the total number of diagnoses in the United Kingdom as of 31 December 2020 to exceed 2.3 million, with more than 40,000 cases having been diagnosed in a single day. The number of people cured was only 5,000, which was much lower than the number of people diagnosed in the same period. At the same time, more than 200,000 out of 2.3 million cases were cured in France.

It can be said that the United Kingdom became the biggest loser in the fight against the pandemic across European countries, and its government's performance in the fight against COVID-19 was another visible failure outside the United States.

8.5 The Division of the Pandemic from the Perspective of Chinese and Western Cultures

The global COVID-19 pandemic resulted in the governments of various countries implementing various policies.

All provinces in China initiated the first-level response to health incidents. The masks were sold out before the pandemic spread. People consciously stayed at home, which enabled China to effectively contain the further spread of the epidemic in just two months. However, when the pandemic first broke out, Westerners, who were oblivious to the dangers ahead, could still insist on not wearing masks. An Italian member of parliament was even ridiculed by a group of members who wore a mask in Parliament and angrily threw a microphone. In the West, there were also narratives such as "it is just a pandemic flu" and "the fatality rate is not high, I will not die," even though these remarks were unbelievable in the eyes of the Chinese people.

There seems to be another way of understanding this in terms of a society's acceptance. If China lost many lives and paid a heavy price due to the pandemic, would its population be able to accept it?

However, faced with the huge losses and casualties due to the pandemic, European and American societies were able to bear such significant losses of lives and a wide range of infections. The support for their governments continued rising, with limited criticisms against them by the people. Some netizens joked that "the first time I deeply felt the difference between China and the West was because of a virus."

8.5.1 Long history behind the differences in mindsets

The difference phenomenon can be attributed to the difference between Chinese and Western mindsets. Human thinking is a function of the human brain that develops to an advanced stage with production and practice and is the product of long-term development in human history.

A nation condenses its long-term knowledge of reality into experience and habits, uses language to form ideas, and forms a mindset unique to the nation. This mindset is a bridge to communication between culture and language.

On the one hand, the mindset is embodied in every aspect of national culture, including material, institutional, behavioral, spiritual, and cultural communication, especially in philosophy, science and technology, literature, aesthetics, art, religion, politics, law, and other production and life practices. The difference in mindset is an important reason for cultural differences. On the other hand, it is also the deep mechanism of generational language and development, which in turn promotes the formation and development of mindsets.

The fundamentals of the differences between Chinese and Western mindsets must be traced to their traditions. It is generally believed that Eastern philosophy originated from the ideas of Confucianism and Taoism, while Western philosophy originated from ancient Greek philosophy. Chinese philosophy focuses on practical social order, which is based on empirical thinking founded on spontaneous empirical forms, while Western philosophy aims at the pursuit of pure knowledge and rational thinking built on pure language analysis.

Therefore, there is a big difference in the understanding of knowledge between Chinese and Western philosophy. Chinese philosophy focuses on practice. There is a saying that "the superior man is modest in his speech, but exceeds in his actions," which refers to the purpose of learning not being knowledge itself, but rather the how one guides their actions. This is classical pragmatic thinking.

On the contrary, from Plato to Aristotle, philosophers regarded sensory perception as knowledge, but only recognized knowledge acquired by reason, and believed that "idea" is the only truth that philosophy should care about. In other words, Western philosophy focuses on "theoretical" pure knowledge, which can only be obtained through language analysis.

In addition, the thinking styles of Chinese and Western communities are perpetual and profoundly affected by economic systems.

China's traditional economy is a typical self-sufficient natural economy, which came into private ownership after the Spring and Autumn Period and Warring States Period. The natural economy lacked contact with the outside world, causing people to limit their vision and thinking. In this kind of agricultural society, people realized that a good harvest cannot be separated from the good weather, and survival cannot be separated from the gift of nature, and they have realized the idea of "all things are one" and "man and nature are one."

Unlike the West, which takes nature as the object of knowledge, it humanizes nature or naturalizes humans, so that the object of thinking is directed to oneself rather than nature. Therefore, traditional Chinese thinking focuses on introverted self-seeking.

Under the feudal monarchy, the political system of "the family and country share common structural features" was implemented, and the family ethics and ethics were extrapolated as the country's ruling order. This concept, which was not ideal at upholding individual rights, also made it easy for people to obey certain specific authorities in group actions. This "agricultural civilized character" has also made the East Asians pay attention to ethics and morality, and seek common ground and stability as the principle of life.

It is because of the huge cultural differences between the East and the West that the pandemic began to spread. The Chinese government used powerful means to effect social isolation, and the tradition of seeking common ground facilitated the adherence to orders in the prevention and control of the pandemic, as well as the successful execution of tasks. With strong cohesion among the people, both materials and personnel can be assembled and deployed quickly.

The Eastern tradition of seeking stability is embodied in the prevention and control of the epidemic. It is mainly manifested in that people cherish their lives more, remain indoors, consciously isolate themselves, and protect personal property and life.

The Western economy originated from the handicrafts, commerce, and navigation industries in the Greek peninsula and its nearby coastal areas. This also aroused the strong interest of ancient Greek philosophers in astronomy, meteorology, geometry, physics, and mathematics, and gradually formed the western scientific tradition of focusing on exploring the mysteries of nature.

The development of the handicraft industry resulted in Western society paying increasing attention to the analysis of processing procedures and skills in tandem with the times. In modern times, Western experimental science developed rapidly, and its adaptability to ideas has been highly empirical, especially since the Industrial Revolution, where "fair theory," "self-actualization theory" and "competitive spirit" were typical characteristics of Western thinking due to the unique organizational, scientific, and democratic edification of the large-scale industrial production. This "industrial civilized character" created a strong spirit of struggle and legal awareness of safeguarding own interests in Westerners, and this adhered to the principles of independence, freedom, and equality.

In this context, it is relatively easy to understand the incredible operations of the United States at the beginning. Trump first opposed the lockdown, then declared that he planned to restart the US economy before Easter, and hinted that he would fire Fauci, the US health expert. The economy was protected over the people. As a result, the economy did not save the people, and it was in danger. In the end, he had to take care of both sides and hurriedly overturned his earlier decision.

8.5.2 Individualism and Collectivism

Among the differences between Chinese and Western thinking, the more interesting one is the difference in the distribution of the two concepts of "individualism" and "collectivism." The study found that there are more individualists among Westerners, and more collectivists among Asians from countries such as India, Japan, and China.

In many cases, the difference between both parties is evident. When asked about attitudes and behaviors, people living in a more individualistic Western society tend to place personal success above collective achievement, which in turn encourages people to seek more personal respect and happiness.

However, this desire for self-affirmation was also manifested in one's overconfidence. Many experiments proved that experimental participants classified as "people from the West, well-educated, aggressive, rich, and democratically-minded" were more likely to overestimate their abilities. For example, when asked about one's abilities, 94% of American professors claimed that their abilities were "above average."

It was worth mentioning that an important cornerstone of American culture is individualism, albeit the presence of a "collectivism" spirit in American culture. The classic American cultural studies text Habits of the Heart focused on individualism and individual commitment to the group. In the preface of the book, the author pointed out that Tocqueville found himself envying, yet worrying about American individualism when he visited the United States in the 1830s.

Like Tocqueville, these authors were also worried that the potentially destructive nature of American individualism runs the risk of being unable to be controlled by family, religion, and community political participation. This danger would turn citizens into isolated individuals, thereby destroying the conditions on which freedom depended on.

In reality, religious leaders, scholars, and politicians such as Winthrop, Tocqueville, and Lincoln, before and after the founding of the United States, as well as contemporary researchers such as Bellah, had a deep understanding of the potential harm that individualism may cause to groups. They were worried that American democracy and freedom would be undermined by extreme individualism.

And what they were worried about happened during this pandemic. As long as many Americans were too self-centered and emphasized individualism and personal interests too much, the epidemic would continue and cause greater harm. For example, on the issue of wearing masks, a public opinion survey found that many American men believe that wearing masks was a sign of "weakness," and they completely ignored the fact that they would transmit the virus to others if they were asymptomatic. Trump's refusal to wear a mask for a long time was also a manifestation of this sort of thinking.

As a result, the tragic pandemic made people see the shortcomings of individualism. The West talks about individualism or collectivism, isolating people and thus respecting individual life choices. In fact, for the founders of Western countries, the most important thing about individualism was not the protection of individual rights, but the high emphasis on personal growth, the social cruel experience of individuals, and the individual's responsibility for the consequences of their actions as the ultimate responsibility.

Whether a person's life is good or not is, first of all, a personal responsibility. You cannot blame God, your parents, and the national government. Therefore, in the United States, the pandemic is a natural disaster. As long as the government uses whatever means it can to help individuals tide over the crisis, the people can accept it. This was why the measures were considered weak by the Chinese, but they brought Trump public support, which was steadily increasing.

Due to the epidemic, Western practices had confirmed the weakness of individualism, which meant that individuals could not take responsibility for social structural problems. If common personal life disasters were caused by social structural problems, letting individuals take responsibility would be a cover-up.

Faced against the pandemic, Western prevention and control measures had not addressed the societal structural issues. They continued using a market-oriented, gradual, and simple expansion of the common disease treatment model to solve some or even most of the social structural issues. Such issues, for instance, high treatment costs and social inequality that result in differential medical treatment, were pushed onto individuals.

In the end, this one-sided development of individualism can only bring catastrophic consequences. In contrast, East Asians who are more inclined to collectivism are characterized by a greater degree of obedience and respect for collective behavior.

China's cultural tradition also determined that China is a more collectivist nation in the world. Studies had found that individuals with high collectivism tend to have lower xenophobic tendencies when facing viruses with higher infection rates, and xenophobia is one of the components of the psychological threat response.

This showed that high collectivists have a lower sense of psychological threat or behavioral response when facing the virus, which meant that the cultural tendency of collectivism has improved people's perceived protective efficacy. Collectivists would perceive that collectivism can protect their safety, although this protection may be their imagination rather than the actual situation.

Collectivism buffers and reduces the impact of susceptibility on xenophobic reactions, and makes people have a higher sense of psychological security and stability, which plays an extremely important role in people's mental health under the pandemic.

Collectivism also makes it easier for people to unite and act. When the pandemic broke out in China, some people were crying, some were saying goodbye, while others were going against the tide. As the world's second-largest economy, organizations were ordered to suspend work, 1.4 billion people were isolated at home, 345 national medical teams and 42,600 medical staff rushed to Wuhan from across China, 19 provinces were supporting 16 cities, prefectures, and county-level cities in Hubei Province except for Wuhan City and necessities, including "medicine boxes," "rice bags" and "vegetable baskets" were dispatched to the people of Hubei and Wuhan in a unified manner.

Wuhan quickly requisitioned several hospitals and arranged more than 3,000 hospital beds. Two hospitals, Huoshenshan and Leishenshan, were built in more than 10 days, and several shelter hospitals were quickly completed. National special scientific research forces concentrated on research and quickly detected the entire virus genome, developed diagnostic kits, selected therapeutic drugs on a major scale, and multiple technical routes for vaccine research and development were carried out simultaneously. The Department of Finance continued to increase funding guarantees for pandemic prevention and control to ensure medical treatment and funds, while state enterprises swung into combat mode and started the production of medical supplies, resulting in the production of millions of urgently needed medical supplies that were increasing more than tenfold within half a month.

From the frontline doctors, who spared no efforts to rescue patients, grassroots staff in various places on standby, ordinary folks who helped China to overcome the battle in their ways, helped China through the storm, to the donation of materials through ground-up efforts to the forwarding of assistance required via the Internet. WHO Director-General Tedros Adhanom Ghebreyesus, commented, "The speediness and scale of China's actions are rare in the world. People are full of praises for China's speed, scale, and efficiency. People highly appreciate this. This is the advantage of the Chinese system, and its relevant experience is worth emulating for other countries." At the same time, the nationwide battle against the pandemic and mutual assistance has also become a vivid portrayal of collectivism.

However, the differences from thinking to cultural politics are to be recognized in the epidemic. The pandemic knows no borders, and mankind has long existed on the earth as a community. Regardless of which civilization is harmed, East Asians or Westerners, it will eventually cause harm to human civilization.

9

Coronavirus Ringing the Alarm Bells

9.1 In the Eyes of the Coronavirus, Everyone is Equal

Like any past major infectious disease, COVID-19 is not only a medical event but also an economic and political one. In addition, it is a communication event.

During the pandemic, it was no surprise that we received more information about it. The rapid development and popularization of the Internet in recent years has enabled information to spread rapidly and provide false information through this medium. From Facebook to Weibo, the misinformation shared is all-encompassing, which included the cause of the pandemic, treatment methods, origin of the virus, to combating it with the double coptis chinensis plant.

As a result, countries had to establish a mechanism to verify and combat false news during the pandemic. A webpage by the World Health Organization is updated regularly to debunk false information.

Interestingly, the unexpected "everyone is equal" approach was achieved in the face of false information. Scholars with an accomplished academic background were not immune to false information either. After news broke that double coptis plant can resist disease, inhibit bacteria, and had preventive effects on COVID-19, it was wiped out at all major pharmacies. The belief that it would be effective against COVID-19 was akin to Trump's ridiculous claim that ingesting disinfectants were effective in combating the pandemic.

In March 2020, a poll conducted by YouGov and The Economist found that 13% of Americans believed that COVID-19 was a hoax, while up to 49% believed that the pandemic may be caused by humans. Although we expect strong judgment or knowledge to help distinguish truth from fiction, many educated people still fell into the trap of misinformation.

Although we knew that false information could cause great harm to society, we still unavoidably believe in false information. Even the so-called "smart people" were not exempted. Why did this happen?

9.1.1 False Information Becomes Real after Processing

False information is considered to be a rumor or public opinion lacking in factual information, or one that is unproven and is difficult for the public to ascertain if it is the truth. It is also called a rumor.

The dissemination of false information has three gatekeepers, namely the disseminator (information maker), environmental intermediary, and receiver. The disseminator of false information transmits false information to the environmental intermediary, which in turn transmits it to the recipient. After the recipient receives and processes the false information, he becomes the disseminator of false information and transmits it to the environmental intermediary, and so on.

The dissemination of false information is a social phenomenon and typical psychological behavior of social groups. Studies in social psychology indicate that false information that meets or caters to people's subjective wishes, impressions, or bias is most likely to be believed and willing to be spread by others and may be carried out randomly and processed based on the particular psychological tendencies of the communicator.

To understand why "everyone is equal in the face of rumors," it may be necessary to establish an objective understanding of false information from a psychological behavioral standpoint. Among them, part of the problem stems from the nature of the message itself.

In today's society, we are bombarded with all kinds of information every day, resulting in us often relying on intuition to assess whether the information is true. Communicators of fake news often use some simple techniques to make the information appear "true," which prevents people from using critical thinking to verify the authenticity of the news source, which is akin to "when the mind is flowing, we will follow by nodding."

Australian National University researchers proved that attaching a picture to an article will increase people's trust in its accuracy, even if the picture has nothing to do with its contents. For example, a common picture of a virus appears at the same time as the text of a new treatment method. The picture does not prove the article itself, but it helps us to visualize the general situation. Therefore, we regard this "processing fluency" as a sign that the statement is correct.

For similar reasons, false information often uses descriptive language or vivid personal stories, providing enough familiar events or figures, such as mentioning the name of a recognized medical institution, so that it can be linked to our prior knowledge, enabling us to feel that the information is convincing.

Even simply repeating a sentence, whether it is the same paragraph or multiple pieces of information, can increase the "authenticity" by increasing its familiarity. We will mistake the sense of familiarity for the authenticity of the event. Therefore, the more we see it in a news feed, the more likely we are to think it is true, even if we are initially skeptical.

9.1.2 Ease of Sharing Exacerbates the Sharing of Information

Apart from the common methods used by people who propagate and peddle false information, the dissemination of information via the Internet also increases the tendency of people being deceived.

When information was disseminated traditionally, a city was realistic, and there was a real relationship between communicators. However, in the Internet age, the city is virtual, and we are more initiated, and these factors involve subtle psychological motives and social attitudes. The convenience of Internet sharing has long overturned the past. We can download and share any content we want or wish to share. The low cost of sharing also reduces the threshold for the dissemination of false information.

Gordon Pennycook, the lead researcher of misleading psychology at the University of Regina in Canada, experimented on the spread of false information. The experiment required participants to determine whether the news headline about the coronavirus outbreak was true or false. When participants were asked to judge the accuracy of the statement, only 25% believed that the false title was true. However, when asked if they would share this headline, about 35% of people said they would share fake news, 10% more than before.

In usual circumstances, before sharing a piece of information, we will assess whether it will be liked and not its accuracy. Pennycook also said, "Social media does not incentivize the sharing of real articles, it inspires participation."

For some shared articles, if the sharer can assess the authenticity of the information, it is possible to determine if it is true or false. Others might transfer the responsibility of assessing to others. When many people share information, they will add a disclaimer, such as "I don't know if it is true, but..."

People take it for granted that if the information is true, it will be helpful to friends and followers. If it is not true, it will still be harmless. Therefore, the potential harm of sharing is disregarded.

In fact, whether it is a promise of self-made remedies or a deliberate cover-up by the government, a promise that evokes a strong reaction among followers will distract people from the issue of authenticity. Of course, the question should be, is this true?

9.1.3 Emotional, False Information is Easier to Spread

If we pay attention to the nationwide dissemination of false information, it is not difficult to find that false information is tagged along with the disseminator's emotions. Finally, it becomes a public opinion that is all-encompassing.

On 31 January 2020, a reporter understood from the Shanghai Institute of Materia Medica, Chinese Academy of Sciences that the double coptis plant, when consumed orally in liquid form, can inhibit COVID-19 and is one of the most effective broad-spectrum antiviral drugs to date. While many found it unbelievable, they disseminated the information and flocked to the pharmacies.

Indeed, emotions are an important aspect of human psychology. Attempting to regulate and control emotions is a common phenomenon in our daily lives. Emotions affect the daily behavior of individuals, as well as their information dissemination behavior. Clore and Schwarz put forward the equivalence theory of emotional information, believing that emotion can be used as an information clue to directly affect decision-making judgments. Interpersonal information dissemination can be regarded as the language expression of individual inner activities. When individuals have strong emotions, they often need a channel to vent, and language is undoubtedly an important method.

The process of information dissemination is a process of both retention and diffusion. For its audience, on the one hand, they will not just store the information, otherwise, the information would not be circulated at all. Usually, we will share the acquired information in the group. However, on the other hand, we will not disseminate all the information.

Studies confirmed that information with vivid language has a greater impact on the behavior of information audiences. Clore and Schwarz's emotional information equivalence theory believes that emotions can be used as information clues to directly affect decision-making judgments. When expressing positive and negative attitudes, emotional expressions make the audience feel stronger attitude tendencies. This means that emotional false information makes them feel that it is easier to understand and more important, which makes them more willing to spread it. In this regard, emotional disinformation can spread faster and more widely.

Therefore, when the disseminator of false information uses emotional language, the audience will observe and experience emotions, which will, in turn, make them imagine an environment similar to these emotions to invoke such emotions, and further spread the information without reservation.

Under the influence of emotions, there might be reports about "Banlangen being able to prevent the virus" again. Without much thought, everyone would flock to the pharmacy again even when they doubt it.

To prevent ourselves from falling prey to false information, we need to change our thinking. We should remain skeptical while trying to find the truth. At the same time, we are to state the facts as simply as possible to combat false information. For example, the facts can be presented with the aid of auxiliary tools, such as images and charts, to enable us to visualize them in the simplest form possible.

If possible, try to avoid repeating the lie itself. Repetition makes the statement feel more familiar and increases the "realness" feeling. Of course, this is not always possible, but we can try at the very least to make the real event more prominent and memorable than the lie, so it is easier for it to stay in our minds.

When discussing our online behavior, we need to detach our emotions from the content and assess its factuality before delivering it. Is it anecdotal or scientific evidence? Can the source be traced? How does it compare with existing data? Does the author rely on common logical fallacies to prove their point? We must do consider the factual basis before sharing it.

There is no elixir, which is akin to people trying to control the virus itself. Only a multi-pronged approach can truly combat false news that may endanger lives. At the same time, it is everyone's responsibility to prevent the spread of false information.

9.2 Social Vulnerability During the Pandemic

The rapid development of science and technology has enabled us to witness and profoundly experience the modernization of society. The duality of modernization includes the normal, positive, and strong sides of society, but it also inevitably has its negative, pessimistic, and fragile sides.

During the pandemic, social uncertainty is increasing unabated. Different problems are constantly emerging in various fields, from finance, education, medical care to employment, which further highlight the fragility of society.

At the same time, the "anti-fragile," or the opposite of "fragile" are increasingly valued by society. Taleb, the father of the black swan, pointed out in the book Anti-fragile that when exposed to volatility, stochasticity, chaos and pressure, risk and uncertainty, anti-fragile things can not only be protected from shocks, but also benefit from them, and establish a mechanism beyond "strong resilience" amidst uncertainty, and to be able to act, self-reform and self-evolve according to the situation.

From "fragility" to "anti-fragility," this is a big test during the pandemic. How do we build a society's anti-fragility system to enhance social resilience and enable it to continue evolving itself and benefiting from various chaos, pressures, and shocks? This has also become a question worth pondering in the post-pandemic era.

9.2.1 Why Does Modern Society Remain Fragile?

The concept of "vulnerability" originated from the analysis of natural disasters in the 1970s. In 1979, the United Nations Disaster Reduction Organization (UNDRO) published a report entitled Vulnerability. Since then, this concept has gradually expanded to the fields of sociology, ecological environment, poverty, and sustainable development, covering natural sciences, medical sciences, computer sciences, engineering sciences, and social sciences, and became a research perspective and analysis method.

Social vulnerability covers all aspects of social life. From a food security point of view, the World Food Program believes that vulnerability is the difference between food security risks and the ability to withstand risks. From an income and health perspective, the World Bank defined vulnerability in the World Development Report 2000/2001 as "a family or individual that experiences the risk of income or health poverty over a period of time, but vulnerability also means that it may face some other risks (violence, crime, natural disasters, and dropouts)."

The Hyogo Action Plan 2005–2015: Strengthening the Resilience of Nations and Communities to Disasters believes that "vulnerability" is "a condition determined by natural, social, economic and environmental factors or activities. As a result of this condition, a community is more likely to be affected by hazards".

From agricultural to industrial civilization, social fragility has its manifestations. Machine civilization has given new characteristics in line with the times. However, due to the systematicity, complexity, magnification, and acceleration of modernization, it is accompanied by the fragility of society.

First of all, systematicity can be divided into holistic aspects, structure, interconnectivity, orderliness, and dynamics. The most important are the integral and indivisible features. Therefore, modern society often affects everything. Secondly, the modernization process is the evolution process of the complex paradigm, and the core of this paradigm is the diversification of self-organizing subjects and the intersection of self-organizing subjects. Hence, going beyond experience and rejecting central control has led to social fragility. Thirdly, the magnification process of modernization is a process of crossing people and society and different systems, multi-infiltration and overflow, and the transmission from one system to another, forming a process of magnification mechanism.

Lastly, there has been an exponential increase in material, technology, and spiritual creation under globalization, which can be seen from Moore's Law and Metcalfe's Law. Mankind has already bided farewell to the era of change within centuries and has entered an era of change within months or even days.

Therefore, when modernization goes beyond experience and even rationale, there is a gap between mankind's capabilities and the modernization led and promoted by man. This gap results in a lag or even difference in cognition, reaction, choice, and decision-making ability, which intensifies the fragility of modernization.

Obviously, during the pandemic, "social fragility" and "secondary risk" were further magnified. It has severely damaged the basic health services of various countries, causing widespread socio-economic interference and impact on the medical system. Measures such as "lockdowns" and social isolation curb the spread of the pandemic have produced extensive and profound socio-economic consequences.

In addition, international relations have resulted in the complication of the pandemic. Whether it is the circuit breaker of the US stock market or the heavy losses in the aviation industry of various countries, the fear of food shortages and recurrence of the pandemic has brought about new panic, reflecting the fragility of modern society.

The impact of the pandemic has increased the number of vulnerable groups in society. It brought about challenges to businesses, the stock market, oil prices, and all aspects of social life. Airlines, restaurants, shopping malls, tourist attractions also suffered heavy losses. The measures are taken to cope with the pandemic such as staying home, stopping work and production, closing cities, and even the entire country, resulting in an all-encompassing global impact.

Hence, as a result of the pandemic, the non-vulnerable groups, which are more equipped to combat risks in comparison with the poor and marginalized groups, may also become vulnerable groups.

In addition, the vulnerability of socially vulnerable groups has deepened. In terms of a global strategy to fight the epidemic, the spread of the pandemic has had a profound impact on public health and socio-economic conditions and caused varying degrees of damage to vulnerable groups.

The global pandemic is a huge disaster that mankind has not experienced for a century since the Spanish flu in 1918. It enables mankind to fully realize that "vulnerabilities are universal." However, due to the widespread inequality in wealth, power, and status of individuals, coupled with social exclusion and discrimination, the pandemic brings about different levels of risks to different groups, resulting in the differing abilities of different groups to fight risks.

9.2.2 Increasing Risks Faced by Vulnerable Groups

During the pandemic, when the risk is greater than our ability to resist it, the fragilities of the world would be reflected. The "fragilities" brought about by the pandemic and the measures to respond to it overlap with the existing "social vulnerabilities," making the most vulnerable groups face greater risks.

First, as a vulnerable group in society, women are more vulnerable to domestic violence during the pandemic. After the outbreak, violence against women and girls has soared to more than 25% and in some countries, it even doubled. As of 3 April 2020, since the "lockdown" on 17 March 2020, France reported a 30% increase in domestic violence cases, while the cases in Cyprus and Singapore increased by 30% and 33%. Domestic violence cases also experienced an upward trend in Brazil, Canada, Germany, Spain, and the United Kingdom.

In addition, from a global perspective, the economic recession had a greater impact on women. Women generally hold lower positions in the workplace than men. They are mostly employed in sectors most severely affected by the pandemic, such as entertainment, retail, tourism, and small-scale agriculture. They are also mostly engaged in low-paid, unestablished fields lacking in legal and social protection jobs. Coupled with the burden of childbirth and childcare, they are more likely to be unemployed during economic downturns and public health emergencies. It is also more difficult for them to return to the labor market after unemployment. Therefore, the COVID-19 pandemic may put women's progress in economic justice and rights in jeopardy for the next few decades.

Studies showed that women took three times more unpaid care work than men. There was neither a gender balance nor a gender perspective in the global COVID-19 policies. The severe lack of gender and sexual health experts also affected key policies. Under the influence of cross-cutting factors, the multiple vulnerabilities faced by women were revealed.

Secondly, children were one of the most vulnerable groups affected by the COVID-19 pandemic. Although the direct impact of the pandemic on children's health seemed to be minor, it was hard to deny that children's lives were severely disrupted during the pandemic, and the risks they faced were reflected in education, food, safety, and health among other aspects.

The issue of children's education amidst the pandemic is omnipresent. According to the United Nations Educational, Scientific and Cultural Organization (UNESCO) data on April 21, 2020, schools in 191 countries were closed, affecting more than 1.5 billion students from preschool to university education. More than 90% of classroom education was also disrupted.

Globally, 50% (826 million) of students do not have a computer at home, 43% (706 million) do not have Internet access at home, and 56 million people cannot use their mobile phones to obtain information because they do not have Internet access. In sub-Saharan Africa, 89% (216 million) students do not have a computer at home, 82% (199 million) do not have a network at home, and 26 million cannot use mobile phones to obtain information. The data gap caused by the global COVID-19 pandemic may cause low-income and disadvantaged students to further lag behind their peers with better conditions.

Third, the vulnerability of the elderly during the pandemic is mainly manifested in their health. An early Chinese study showed that the mortality rate of adolescents under the age of 20 was very low. As we age, estimates of severity were reflected in case reports, with an average age between 50 and 60 years old. According to a research report by the Robert Koch Institute, a German disease control agency, as of 2 May 2020, 67% of those infected were 15–59 years old, while 87% of those who died were aged 70 years old and above, but they only accounted for 19% of infections in this age group.

On the other hand, there were reports that in the case of extreme lack of medical resources, some Italian doctors had made it clear that they would abandon the "first come, first served" principle and leave the ventilators for young patients who are more likely to survive. This reflected the intersectional discrimination against the elderly, which led to a humanitarian crisis, causing more controversy.

Fourthly, people with disabilities, as a special group, are inherently restricted in their access to education, health care, and income opportunities and their participation scope in community activities, including being more likely to live in poverty and experiencing violence, neglect, and abuse. The fragility of the environment during the pandemic also exacerbated the situation. Due to the lack of available public health information for people with disabilities, it has been difficult to obtain sanitation facilities. Therefore, after contracting COVID-19, they were more likely to have serious health conditions.

In addition to the prominent vulnerabilities of women, children, the elderly, and the disabled, the impact of a global pandemic on refugees and migrants was disastrous.

During the pandemic, large numbers of refugees were forced to leave their homes due to conflicts, violence, and other reasons. They lived in refugee camps and other uninhabitable environments and had difficulty obtaining clean drinking water, sanitation systems, and health care facilities. Many people were not able to ensure their basic living conditions, much less talk about frequent hand washing, personal protection, and social distancing.

More than 85% of refugees and almost all domestically displaced persons in the world live in low- and middle-income countries, and the spread of the pandemic had a severe economic impact on them. Starting from March 2020, countries began implementing lockdowns and other public health measures. By early April, UNHCR and its partners have received more than 350,000 calls from refugees and internally displaced persons from the Middle East and North Africa alone. Most of them requested emergency financial assistance to meet their daily survival needs.

There were immigrants apart from refugees. According to the FBI report, after the COVID-19 outbreak occurred in Wuhan, China, hate crimes and harassment incidents against Asian Americans increased, especially after Trump labeled the pandemic as a "Chinese Virus." It amplified xenophobia and aggravated the stigmatization of Asian communities and hate crimes in the United States and abroad. There have been incidents of discrimination and attacks on Asians in many countries, including Germany, the United Kingdom, the Netherlands, and Australia. According to United Nations reports, COVID-19 was considered a "foreign" disease, and immigrants of Asian and European ancestry were stigmatized as spreaders of the coronavirus.

With a focus on socially vulnerable groups, social vulnerability is concerned with the impact of the internal characteristics of the social system on vulnerability, which are poverty, inequality, marginalization, social deprivation, and social deprivation caused by the inherent instability and sensitivity of the social system. In addition, social vulnerability is dynamic and relative, which manifests differently in different situations. Income, health, risk, sensitivity, exposure, environment, and social relations are all key factors affecting vulnerability.

The social vulnerability assessment was not only an assessment of social vulnerability but also a reflection of social resilience. Therefore, social vulnerability assessment will also become an important entry point to realize the combination of vulnerability and resilience research. From "fragility" to "anti-fragility," this was a big test during the epidemic. How we can build a social anti-fragile system to enhance social resilience is also a question worth thinking about.

9.3 The Increasing Gap Between the Rich and Poor After the Pandemic – From the Perspective of The Luxury Goods Market Recovery

A global pandemic did not appear to be the best time to sell expensive luxury goods, especially when the global economy was expected to shrink by 5% in 2020.

But in fact, signs of economic recovery within China were obvious as the pandemic was brought under control. Among them, luxury goods, which recovered by 120% in June, enjoyed the fastest recovery. Data showed that most luxury brands reached the levels in April 2019. In May, the growth rate of luxury goods reached 20% to 40%. It was reported that Beijing SKP's May Day sales tripled last year, regardless of whether it was Chanel or LV. There were still queues at 9 p.m.

In addition, some luxury brands also released plans to increase prices in May. On 13 May, Chanel announced that they planned to increase the prices of handbags and small leather goods worldwide, between 5% and 17%, due to the impact of the pandemic. Louis Vuitton raised prices again in May after raising prices in March. This meant some luxury niche markets were still booming, with some asset prices soaring in unexpected ways, despite the economic difficulties caused by the pandemic.

On 2 August, Christie's auction house sold another pair of sneakers worn by Michael Jordan at Gottahaverockandroll for US$474,696 (US$379,757 on top of a 25% buyer's premium), surpassing the previous two. This was a new record, which could throw billionaires off guard as well. In addition, there were 12 pairs of Jordan sneakers, which would be auctioned at two different auction houses over the next 10 days. One pair was expected to sell for between USD 350,000 and USD 550,000, while the other pair was estimated to be sold for USD 650,000 to USD 850,000 according to Christie's auction house.

Whether it was the soaring asset prices or the online auction of luxury goods, they reflected the effect of exports on domestic sales, with the younger generation becoming the main force in luxury sales. This also emphasized the fact that the pandemic did not only give rise to many losers, but also many winners, some of whom were still prepared to spend money.

9.3.1 How Did the Pandemic Worsen the Gap?

The academic circles had various discussions about the shape of the letters concerning the trajectory of the future economic recovery. As of August, the market situation can be called a "K-shaped recovery" at best. Although people panicked in March, different groups of people began having two very different experiences.

Firstly, the pandemic increased the wage gap between skilled and unskilled workers. Although every worker was affected, the impact was greater on unskilled workers more than skilled workers. One reason is that it was more difficult to replace skilled workers than unskilled workers. Hence, unskilled workers were more likely to be laid off or have their wages reduced during the pandemic.

At the same time, high-skilled occupations such as managers were more suited for work from home arrangements than low-skilled occupations such as manual operations. For the rich and those who can work from home, their lives continued amidst the inconvenience caused by the pandemic. They just needed to face new technologies and new habits, but generally speaking, their lives were not significantly impacted as a result of the pandemic.

Secondly, the pandemic decreased the share of labor income and increased the share of capital income in national income. In general, the various pandemic control measures had a lesser impact on capital-intensive industries such as manufacturing as compared to labor-intensive industries such as service industries. This is because the former were mainly workers operating machines, while the latter required employees to directly serve customers.

Research has proven that the impact of COVID-19 was the greatest on service industries such as retail, tourism, hotels, entertainment, and leisure. For many large global companies, the pandemic was just a bump on the road to occupy a larger market position. Due to unprecedented central bank liquidity measures and investors' enthusiasm, their stock prices not only rebounded but reached new highs, especially so for technology companies. In addition, for large companies, the conditions for their access to the credit market had never been more favorable. Amazon broke the record for the lowest interest rate for US corporate bond issuance in June.

Lastly, the pandemic's impact was greater on small, medium, and micro-enterprises, and vulnerable groups such as women and the elderly. Compared with large enterprises, the impact of economic contraction on small, medium, and micro-enterprises will be much greater. As the days go by, more physical stores and restaurants will close down, and digital transformation will remain challenging for the F&B and retail industries. In addition, these F&B and retail industries, which were severely hit by the pandemic, happen to be industries with a high concentration of low-income workers, who are more involved in the short-term work programs and may even face the threat of unemployment.

Women are more likely to work in labor-intensive industries, which are more affected by the pandemic, for instance, the service industry. The elderly is the most vulnerable to infections during the pandemic, so they are the most vulnerable, and their income is much lower than that of the working population. Lastly, during the pandemic, workers who worked at the frontlines and transported food and medicine faced a higher risk of infection, and most of these were holding positions with lower wages.

9.3.2 Everyone's Equal, But not so Equal

According to the Asian Development Bank analysis, if the pandemic continues for 6 months, the GDP of the Asia-Pacific region would fall by 9.3%, which is equivalent to a loss of 2.5 trillion US dollars. Employment would be reduced by 170 million, while the number of the poor would increase by 140 million. The longer the pandemic, the greater the impact.

Apart from causing economic contraction and increasing the number of poor people, COVID-19 will also exacerbate income disparity.

Some researchers have calculated the number and proportion of low-income groups in China in 2019 based on the 2018 household survey data, assuming that the income distribution of residents remains unchanged during the two years. According to the annual household income of 100,000-*yuan*, low-income groups in urban and rural areas account for nearly 65% of the national population, equivalent to 900 million people. If according to the relative standard (2/3 of the median income), then the proportion of low-income people is about 37%, about 510 million people. This also means that China is still a society dominated by low-income people, which is also a basic characteristic of developing countries.

A wealth distribution research report issued by the German Economic Research Institute DIW showed that in terms of total net wealth, the richest 0.1% of Germany own 20% of the total net wealth, with the richest 1% accounting for 35.3%, while the richest 10% accounts for approximately 67%. This means that more and more wealth is concentrated in a smaller group of people. The research report also showed that people's happiness and satisfaction index is directly proportional to wealth.

This set of numbers and results are particularly eye-catching in the face of social groups who are running around for their jobs and are anxious during the pandemic. In the face of disease, it seems that everyone is equal, and even the Prime Minister of a country cannot escape from the fate of being infected with COVID-19. However, it seemed difficult for everything to be equal. The conditions for epidemic prevention differ for everyone, and in the face of a life crisis, our resource reserves are also different. Different people are developing in different directions.

There is no doubt that the increase in the income gap will intensify social conflicts and curb consumption, thereby increasing the risk of economic recovery. Generally, when inequality increases, it will take a while to observe if there is an increase in social separation.

The research of Stanford University economist Raj Chedi showed that intergenerational mobility varied greatly due to the different communities in which individuals grew up. Chedi found that for every year children live in the most upwardly mobile communities, their adult income levels will therefore be 0.8% higher than the national average. Also, every year of living in the least upwardly mobile community will lead to a 0.7% reduction in children's income as an adult.

In response to the widening income gap, many Asian countries, including China, India, and Southeast Asian countries, have elevated the promotion of inclusive economic growth as a basic national policy in recent years. For example, China began implementing the Western Development Strategy to narrow the regional gap in 2001. In 2006, it put forth the idea of building a harmonious society. In 2015, it ensured the elimination of extreme poverty by 2020 as an important indicator for building a well-off society in an all-rounded way. At thc same time, the income gap was narrowed by reforming the household registration system, raising the minimum wage, and expanding the coverage of rural social security. In 2020, China proposed the anti-monopoly policy against Internet companies, as well as the curbing of disorderly expansion of capital. These were effective ways to control the rapid expansion of the gap between the rich and the poor in society.

However, society still faces the bifurcation of the two paths of "K-shaped" recovery even after the pandemic. As many public support projects are coming to an end, financial market trends reflect expectations of a "V-shaped" recovery, and it is wise for policymakers to think about the current gaps. Although effective policy intervention cannot completely prevent this from happening, it can greatly reduce its impact at the very least. It is only by dealing with the negative consequences of the "K-shaped" recovery can we avoid the "L-shaped" recovery that everyone may face.

9.4 Humanity Crisis During the Pandemic

During the COVID-19 pandemic, Yuval Harari, author of A Brief History of Mankind, published an article The World After Coronavirus in the Financial Times. He discussed the pandemic, which we are experiencing, and where the world is heading after the pandemic.

The continuation of the global pandemic has undoubtedly put mankind in crisis. It changed the way people live, which only affects the healthcare system, but also the economy, politics, and culture. Apart from the medical, economic, political, and cultural crises, the real crisis lies with mankind.

9.4.1 Humanitarian Crisis During the Pandemic

Camus told four stories about World War II to explain the crisis of mankind in his speech at Columbia University.

Two prisoners were found in a large city in Europe. After a night of severe torture, they were still bleeding. The gatekeeper of the building rearranged everything in order as if nothing had happened. When one of the prisoners reprimanded her of her attitude, she angrily replied, "I never care about the tenants."

In Lyon, a prisoner was taken out of his cell to be tried by a tertiary court. During a previous interrogation, his ears were torn apart. The German officer, who led him out, asked him in a sympathetic and caring tone, "How are your ears today?"

In Greece, a German military officer was preparing to execute three brothers who were taken hostage. Their old mother begged him to let her son go. He agreed to pardon one, but only if she chose to take one of them. As she could not decide, the soldiers raised their guns. At the last moment, the mother chose the eldest son because he had a family to take care of. But in doing so, she also sentenced her other two sons to death. And this was exactly what the Germans want.

A group of exiled people was sent back to France via Switzerland. They saw a funeral procession when they were about to enter the Swiss land and burst into hysterical laughter. They shouted, "This is how the dead are sent here."

Does the crisis of mankind exist? Camus shared: "As long there are people in this world, there will be those who are indifferent, those who are in hypocritical friendships, as well as those who look upon death and punishment with curiosity or without any response. There will also be those who leave others to fend for their lives, and yet not feel horrible or shameful. As long as human suffering is seen as a troublesome form of hard labor, akin to the exhaustion from queuing to get a little butter, it can be concluded that a mankind crisis exists. "And this kind of mankind crisis, from individuals to society, is profoundly experienced in the pandemic.

The pandemic caused a global humanitarian crisis. Countless people got sick every day, and countless lives were lost. The surge in the number of infected people has caused a "run" on medical resources and supplies, and the increasing demand for medical supplies and professionals placed heavy pressure on the medical and health systems and security systems of various countries. During the pandemic, a doctor in the emergency department in Madrid revealed to the New York Times: "We must choose the person to be intubated as it can no longer be extended to everyone." Spanish 20 Minutes reported that critically ill patients over 60 years old cannot enter the ICU ward.

In the United States, the elderly, a vulnerable group, was further weakened and marginalized due to age discrimination, and their right to life cannot be guaranteed during the pandemic. On 23 March and 22 April 2020, Texas Lieutenant Governor Dan Patrick stated twice in an interview with Fox News that he "would rather die than see public health measures harm the U.S. economy" and agreed to take the "risk to restart the American economy" at the expense of the lives of the elderly.

On the other hand, there were still differences between national isolationism and global unity. As to whether the world can unite to defeat the virus, firstly, it needs to share information on a global scale, but the information barriers of various countries still exist. Faced with the huge threat to the universal right to life and health on a global scale, the U.S. government, instead of devoting its energy to the prevention and control of the pandemic, instead wielded a hegemonic stick to stir up trouble, try to divert attention, shirk responsibility, and unite and cooperate with the international community to deal with the pandemic, resulting in serious destruction.

Also, Iran, as one of the countries with the highest infection rate in the world under the epidemic and a close ally of Britain, the United States, and other countries, was calling on the Trump administration to relax sanctions on Iran and allow 80 million people to ship medical supplies and humanitarian assistance to Iran. However, some U.S. leaders regarded COVID-19 as a tool for "extreme pressure," ignoring the large number of innocent civilians who may die as a result, further revealing the humanitarian politics in the United States.

During the pandemic, many flights were canceled, international students were stranded, and online requests seeking assistance were overwhelming. This was not only about universal risks and survival conditions, but also about different systems, policy logic, separated groups of people, and unequal life crises. Also, due to various reasons such as race, nationality, cultural affiliation, policy preferences, and disproportionate information, foreign students stuck between border controls and conditional mobility became "homeless" during the epidemic. Returnees, akin to refugees in a biological crisis.

In addition, the pandemic exacerbated the inequalities that individuals and families face in humanitarian crises. As governments focus on their respective countries and consider their citizens first, others in need of humanitarian assistance are ignored. With the spread of COVID-19, governments have implemented travel restrictions, which inadvertently blocked the transportation of rescuers and hindered humanitarian response operations.

In some cases, travel was imposed because the government aimed to protect its citizens. Therefore, rescuers already in the country cannot provide vital services. In Greece, even though the rest of the country has returned to normal, asylum seekers and immigrants were still subject to strict isolation and blockade policies, which limited their access to basic services. According to Amnesty International, some refugee camps across the globe faced a greater risk of hunger than the threat posed by the virus itself due to lack of access to aid.

The interruption of rescue operations meant that we have lesser access to soap and water, which is essential to control the spread of the pandemic. Other non-pharmacological interventions, such as maintaining social distancing, avoiding gatherings, and crowded indoor spaces, did not apply to many countries or regions that require humanitarian assistance. For example, the population density of Cox's Bazar in Bangladesh is 40,000 people per square kilometer, 40 times the population density of the entire country. In this case, it was extremely difficult to isolate confirmed cases. Also, personal protective equipment was often difficult to obtain due to the export restrictions imposed by various countries.

The world has changed multiple times, and it is still changing. Therefore, everyone must adapt to a new way of living, working, and building relationships. Like all changes, some lose more than others, and they will be the ones who would lose too much. The best one can expect is that the depth of this crisis enables mankind to recognize the crisis and break our limitations of cognition. The first step in solving the crisis should be to face the crisis.

9.4.2 Global Public Governance Crisis During the Pandemic

The global spread of COVID-19 was not only due to the increase in globalization and close international personnel exchanges. On the contrary, it was precise because globalization lacked depth and the coordination and cooperation mechanism among various countries (or regions) was not well-established enough to form a global "game of chess" public health governance system, which led to the revealing of many of its limitations.

On the one hand, as the United Nations' specialized agency for health affairs, the World Health Organization (WHO), with 194 member states, is the most authoritative and professional international organization in the field of global public health security. It should play an important role in the early detection and warning of sudden infectious diseases, coordination of prevention and control strategies, sharing of diagnosis and treatment methods, and organization of international assistance.

However, due to problems such as insufficient authority and shortage of funds, WHO's mobilizing of relevant global resources and coordinating the joint efforts of countries (regions) to respond to the pandemic was not effectively executed. For example, some Western countries were nonchalant, placing political opinions above professional authority, and individual interests above public health, resulting in the missing of precious window time and causing irreparable losses to the country and the world.

The World Health Organization issued a warning about COVID-19 to all countries and regions in mid-January and declared it as a public health emergency of international concern on 30 January. However, some Western countries did not pay enough attention and prepare sufficient resources, which resulted in them having to pay a heavy price. Concerning the naming and traceability of the virus, WHO promptly gave it a scientific name, and repeatedly pointed out that there should be no stigmatization and racial discrimination. However, some politicians and media such as the United States disregarded the professional authority of the World Health Organization and acted arbitrarily, seriously damaging the overall situation of global pandemic prevention and control and posing a great threat to the safety and health of mankind.

On the other hand, it is not an exaggeration to say that the COVID-19 pandemic is the most serious global crisis since World War II. Its impact on economies has intensified, bringing instability to countries and regions, and even led to turmoil and conflict for some. This also meant that the international community requires large-scale, coordinated, and comprehensive multilateral response measures. It is only through strengthening international cooperation and building a strong synergy that we can fight the pandemic as one human race.

However, a few Western countries, especially the United States, not only failed to provide necessary assistance and support after COVID-19 broke out in China, but they were also the first to adopt targeted travel restrictions. When the outbreak occurred in their countries, there was a shortage of medical supplies. However, they did not concentrate their efforts to control the pandemic and seek cooperation with China. Instead, they politicized, labeled, and stigmatized COVID-19 by imposing US certification thresholds on China's anti-pandemic drug exports, attempted to degrade the image of Chinese products with US standards and certifications, and launched public opinion wars to discredit China. The United States even initiated compensation lawsuits in an attempt to cover domestic contradictions, which negatively impacted global cooperation in the fight against the pandemic.

At the same time, the rise in unilateralism and populism caused by COVID-19 also became a catalyst for changes in the global order. Firstly, there was a lack of strategic mutual trust among major powers, which caused partiality and "conspiracy theories" to prevail, the politicization of the COVID-19 crisis and some economic and trade issues, as well as the formation of a zero-sum game. The second was the lack of a global coordination mechanism that kept up with the times, and it was difficult to form a global discourse system with the greatest common divisor. It was easy to generate political sentiment and encourage conservatism. Third, there was a lack of theories to guide development in the new era globally, with insufficient analysis of the overall role and related links of various elements of economic and social development, and a lack of systematic and in-depth research on the reform of the global governance system, multilateralism, and the community with a shared future for mankind.

COVID-19 may exacerbate existing problems while creating opportunities that occur once in a century. To seize the opportunity and improve global public health governance, it is necessary to establish a value orientation that conforms to the broadest and most fundamental interests of mankind, which are to put human life and health in the first place and uphold the value of life and health above all.

Among them, the core characteristics of the "community with shared future for mankind" concept are people-centric, respect for human life and health, use of basic attributes distinguishing humans and animals as a link to disregard differences in gender, skin color, race, belief, and country, as well as to encourage mankind to help one another and overcome difficulties together based on the greatest homogeneity between people and universal connection of sharing weal and woe.

9.4.3 Crisis of Scientific Awareness During the Pandemic

In the past, we knew very little about the plague. When facing unknown, highly contagious diseases, disgust and fear resulted in us placing value on religious superstitions and moral meaning.

In ancient times, the plague was used as a tool for sin. Apollo in Iliad punished Agamemnon for abducting Kleis's daughter and infected the Akeans with plague. In Oedipus, the plague swept through the Kingdom of Thebes as a result of the crimes committed by the King. The rat plague of 1348 occurred because the citizens of Florence behaved too unruly as described in Boccaccio's Decameron.

Leprosy was considered a symbol of social corruption and moral corruption in the Middle Ages. The word lépreuse in French, when used to describe eroded stone surfaces, meant "like leprosy." It was seen as a trial of the entire society. By the second half of the nineteenth century, it was common to interpret catastrophic pandemics as moral laxity, political decline, and imperial hatred.

In 1832, Britain linked cholera to alcoholism. The minister of the British Methodist Church claimed, "Anyone who suffers from cholera is an alcoholic." Health became a proof of virtue, just like a disease, which became a proof of depravity.

Modern science confirmed that praying to disperse the virus, killing cats to eradicate the plague, using Banlangen to treat strange diseases, and leveraging remedies for peace of mind, are unscientific pandemic prevention and response measures, which were not only ineffective in containing the pandemic but may also increase its spread and hurt more lives.

Conversely, the development of modern microbiology and Koch's law in the second half of the 19th century enabled many infectious diseases to clarify the pathogens and transmission routes. On the one hand, they provided the scientific basis for targeted pandemic prevention measures and vaccine development. Antibiotics, recombinant vaccines, and antiviral drugs invented since the middle of the 20th century, greatly enriched the means for humans to prevent and treat infectious diseases, and significantly reduced the threat of many infectious diseases to our society.

In addition, the scientific spirit played a huge role. For example, Jenner invented the vaccinia vaccine in 1796 through the scientific spirit of careful observation, rational analysis, and rigorous experimentation. For example, during the cholera outbreak in London, England in 1854, Snow determined a well as the source of infection through map marking, an epidemiological investigation method, and its closure thereafter greatly eased the pandemic.

However, these authoritative medical guidelines and recommendations, which accumulated thousands of years of human wisdom and are respected by thousands of people, are supposedly practices with the strongest evidence in modern medicine. However, they were hit and undermined by the politicization of science during the pandemic.

At the beginning of October 2020, a series of the world's top medical science journals such as *Lancet, Science, Nature, and NEJM,* published articles and severely criticized US President Trump's anti-pandemic performance in an unprecedented manner.

Among them, the *New England Journal of Medicine (NEJM)* published an editorial on October 8 to strongly condemn the incompetence and failure of the Trump administration, which turned COVID-19 into a tragedy. This was the first time in its 208-year history that NEJM had opposed a candidate on the eve of the US general election. It was also extremely rare to publish an editorial listing the names of its editors.

The article pointed out that the United States originally had the world's leading biomedical research system, rich expertise in public health and other aspects, as well as the ability to transform its expertise into new treatments and preventive measures.

However, the Trump administration had largely chosen to ignore or even belittle experts, destroying the credibility of the Centers for Disease Control and Prevention (CDC), marginalizing the National Institutes of Health (NIH), and politicizing the U.S. Food and Drug Administration. It attempted to conceal the truth, promoted the spread of lies, and constantly undermined the people's trust in science and the government.

Academia was disappointed not only in Trump's poor performance but also in his destruction and trampling of scientific consciousness during the epidemic. Speaking about the plague itself, we should scientifically and rationally face diseases including pandemics, establish a more complete disease narrative in medicine, restore the disease to physiology and medicine, and treat the impact of the plague objectively. This will be the most powerful weapon for mankind to defeat the plague.

9.5 Coronavirus Ringing the Alarm Bells

Since COVID-19 was defined as a global pandemic by the World Health Organization, it became the most important variable affecting the global economy and political trends. Its spread and continuity had repeatedly refreshed expectations from all walks of life.

The pandemic's scope and impact on people, the economy and society reached an all-time high in 100 years, with some holding the view that its impact had surpassed the World Wars in some respects. This was because the Wars did not cause economies to shut down, as well as result in almost every country facing crises and challenges.

During such a historical pandemic, discussions about its impact on the economy, society, and global political structure are aplenty. However, such discussions and analyses rarely reflect the relationship between mankind and nature and thus adjust our behavior. The relationship between mankind and nature is the debate between anthropocentrism and non-anthropocentrism. Although the pandemic significantly changed the world, its ultimate impact remains unknown.

From the disorder of the global governance system exposed by the pandemic to the general rise of nationalism and populism, it is inseparable from the discussion of centralism. This will also have a mandatory impact on the reconstruction of our relationship with nature after the pandemic.

9.5.1 From Anthropocentrism to the Anthropocene

Previously, humans and nature were always at the opposite ends of ethics and morals. We had always been eager to conquer nature because our power was too weak compared to nature. Therefore, the ethical choice of mankind was always to try constructing their identity as the conqueror of nature.

The source of anthropocentrism can be traced to Protagoras, a 5th century BC Greek philosopher, who claimed that of all things the measure is Man, of the things that are, that they are, and of things that are not, that they are not. This thought of using man as the basis for the existence of things influenced Plato, leading him to construct a mankind-centered world from the concept of man.

In Aristotle's writings, anthropocentrism was more specific. He stated in Politics that Nature cannot create anything with a purpose and use. Therefore, all animals must be created by nature for humans. In other words, the value of animals is to provide services to mankind, who is the master of animals and not morally obligated to animals.

In the medieval theological system, God not only created man but also created all things centered on man, so man is the master of all things. Ptolemy's "geocentric theory" was still based on "anthropocentrism" and believed that humans were not only at the center of the universe, but also at the center of the universe in the sense of "purpose." Therefore, man is the master of all things in the world. This thought gave birth to the birth of anthropocentrism with theological teleology as the main idea. In the 17th century, French philosopher Descartes took "I think, therefore I am" as the basis of epistemological philosophy, emphasized that the purpose of science is to benefit mankind, and advocated "making oneself the master and ruler of nature with the help of practical philosophy."

In the development and evolution of anthropocentrism, both Bacon and Locke emphasized the power of knowledge and advocated that man should be the master of nature. Both Descartes and Kant emphasized that man should be the ruler of nature and the supreme legislator of nature. They all insisted on taking man as the standard for understanding nature and evaluating the world entirely from man's interests.

After the 18th century, several industrial revolutions promoted the rapid development of science and technology. We chose science to change our weaker position in the natural world and were able to use science and technology to conquer and change nature in certain aspects. We made guns to hunt animals, improved fishing gear to catch fish and shrimp, invented drugs to deal with mosquitoes, collected coal to drive away the cold, felled trees to build houses and built roads to facilitate travel.

At the same time, the powerful abilities possessed by human beings had a significant impact on Earth's ecosystem, which led to the concept of Anthropocene. Klaus Schwab (Klaus Schwab), the founder of the World Economic Forum (WEF) in Davos, said: "We must remember the era in which we live now, which is the 'Anthropocene' or 'Era of Humanity.' For the first time in history, our activities were the main force shaping all life support systems on Earth." In other words, our behavior had almost become an important all-encompassing force affecting Earth.

Now, the concept of Anthropocene is no longer just a concept and viewpoint for discussion. On 21 May 2019, the well-known British scientific magazine Nature reported that the authoritative scientific research group "Anthropocene Working Group" voted to formally apply to the International Geosciences Union to add "Anthropocene" as the geological representative starting from the middle of the 20th century. There was sufficient evidence to show that in the decades after 1950, some of the Earth's geological features had undergone significant changes. The carbon dioxide content in the atmosphere, global temperature, and sea level have changed drastically compared to those in the past 10,000 years.

9.5.2 Ringing of Mother Nature's alarm

To some extent, we had become the masters or the most important rulers of Earth and possessed the power to transcend nature. However, Engels also said: "We should not be overly carried away by our victory over nature. For every such victory, nature will retaliate against us."

After COVID-19 started ravaging around the world, the United Nations Environment Program (UNEP) released a video titled "COVID-19: Nature's alarm bells." In this video, the United Nations Environment Program pointed out that the COVID-19 pandemic is a wake-up call that nature is sounding its alarm bells to mankind. Our destruction of nature has endangered our survival. On average, one new type of infectious disease threatening human health will appear every 4 months. Among them, 75% of new infectious diseases were transmitted from animals to humans."

It is because of the existence of biodiversity that viruses do not spread and get transmitted easily to humans. With the increasing scope of human activities, there has been more contact between humans and wild animals. Coupled with the illegal hunting and trading of wild animals by criminals, viruses are more likely to be transmitted to humans.

COVID-19, SARS coronavirus in 2003, and MERS coronavirus in 2012 belong to the β genus of the coronavirus family and are zoonotic viruses. Medicine has yet to determine the source, cause, and intermediate host of this virus, although studies point to bats as the first carrier of the virus. However, there is still a long way to go for science to prove how bats transmitted the virus to humans.

Currently, about 1.7 million unknown viruses exist on Earth. After they were transmitted to humans, many of them would be like COVID-19, which are unknown to humans, making them difficult for people to be resistant. After the emergence of COVID-19, we expect to find therapeutic drugs and vaccines to control infectious diseases. However, due to the dwindling number of wild animals and plants, our resources to deal with unknown infectious diseases are also declining.

Professor Jean Michel Claverie, a virologist at Aix-Marseille University in France, pointed out that once viruses released from glaciers and frozen soil come into contact with suitable hosts, they may be resurrected. Therefore, if humans were to be in contact with pandemic viruses that were frozen, they may be infected and kickstart a new pandemic.

In August 2016, a 12-year-old boy died after contracting anthrax in the Yamal Peninsula tundra in Siberia located in the polar circle. At least 20 others were admitted to the hospital. One possible origin of this anthrax infection was a reindeer, who was infected with anthrax, died, and its carcass was buried in frozen ground 75 years ago. The frozen ground was melted in the high temperatures during the summer of 2016, thus exposing its carcass, releasing anthrax into the nearby water and soil, and the food chain. As a result, more than 2,000 reindeers grazing nearby were infected, which resulted in human infections.

Of course, whether a virus that was once lethal is released after freezing is still active, or whether its lethality is weakened, requires further academic research and confirmation. However, with frequent occurrences of climate warming, environmental changes had certainly pushed us into a situation where people face greater risks. The increase in uncertainty is closely related to climate change. However, we are still not sufficiently vigilant.

9.5.3 Confirmation of Human Anthropocentrism to Human Subjectivity

It is undeniable that human activities resulted in mankind inching towards danger. The current pandemic had a major impact on us and caused lives to be lost, alongside economic regression. With the emergence of a humanitarian crisis, governments of various countries had taken unprecedented measures to deal with the emergency of the COVID-19 pandemic. With the reconstruction of the economy, the opportunity that we should seize is to take the same drastic action on climate change and realize the transition from anthropocentrism to decentralization.

Decentralization does not negate the dominant position of human beings. On the contrary, it is only by confirming their shift from the center to the dominant position can our responsibility be confirmed, and the way to solve the humanitarian crisis be found. Humans cannot be the center of the universe or nature, but they can be the main body to protect the latter. Humans cannot be the ruler of nature, but they can be the protector of nature. It is difficult for nature to achieve this as an object, be it in the past, present, or future. It is only possible to solve the increasingly serious environmental pollution problem and eliminate the ecological crisis after we ascertain our dominant positions and perform our roles.

In reality, environmental pollution and ecological crises were caused by us, so we are the only ones who can solve the problems we created. While making ethical choices, we would be able to realize that the environment is closely related to our quality of life. We can create a great living environment for ourselves and protect it. However, things are always the opposite in reality, and we have not yet achieved this.

Men have acquired their form through evolution and entered the stage of ethical selection, which refers to distinguishing between right and wrong, good, and evil through teaching, and being a moral person. It also means that human beings must realize the value of their existence as a subject, and the responsibilities and obligations they should bear in solving the ecological crisis, to genuinely choose a scientific path that conforms to human ethics. The correct choice for mankind should be to solve the ecological crisis rather than to create a new one.

In the movie Alien Bacteria, a manmade satellite fell on a small town in Utah. Those who opened it released a deadly virus, which killed almost everyone in the town, leaving behind only four survivors and causing panic across the globe. Just when people were at a loss, an encrypted code passed by humans in the future pointed out that the only nemesis of this virus is the bacteria living on the surface of the mine, which the government wants to mine. Finally, the research team cultivated the bacteria in large quantities and finally eliminated the virus, and saved human beings.

This film seemed to be a metaphor for this era. We had evolved to this level of modern civilization. Yet, it was the lowest level creatures, which eventually saved our civilization. This is more like a warning. If we do not respect nature, viruses would be a natural tool to correct our misbehaviors.

Lao Tzu said in the Tao Te Ching that Heaven and Earth would unite, and sweet dew would fall. People would by themselves find harmony, without being commanded. Heaven is equal, and the sweet dew descended from the harmony of Heaven and Earth, with both people and all things equally moisturized. At the same time, the balance of the heavens is connected with its selflessness. Because of the selflessness of the heavens, it can be well-balanced, and everything can be nourished. In Matthew 5:45 of the Bible, it stated that "for he makes his sun to rise on the evil and the good and sends rain on the just and the unjust."

Whether it is Lao Tzu or the Bible, what they conveyed were the problems that mankind faced for thousands of years, which are in the universe, nature, on Earth, the position of mankind, and other things being equal in the eyes of heaven. Therefore, we need to be in awe of viruses, the environment, wild animals, and nature.

There are pros and cons concerning the existence and occurrence of good and bad things, which is akin to the existence of viruses, which can bring epidemics to humans and eliminate bacteria for humans. "Tao is great. Heaven is great. Earth is great. People are also great. Thus, people constitute one of the four great things of the universe." (Lao Tzu Chapter 25 The universe has four major systems, namely Tao, Heaven, Earth, and People, with people being ordinary members of these four systems. Human beings are neither nobler than the Earth, heaven, and Tao, nor do they have a higher status than things. In the post-pandemic era, the big test for human beings is to learn to treat them with awe, and respect, abide by and cherish the laws of how the Earth, heaven, and Tao operate, and truly realize the harmonious coexistence between man and nature, enabling us to survive and develop for a long time.

The pandemic is causing huge economic and political turmoil unless we can find a suitable solution. The aggregate impact of the pandemic remains unknown. The post-pandemic world will undergo major changes, including to mankind. This is not inevitable as we are still able to make choices. But, the most important thing is that we can always make the right choice.

Afterword

Self-discipline and Nature: A History of Humanity's War Against the Pandemic enabled us to see that plagues recur and exist in various forms through the course of history. It was stated in Ecclesiastes Chapter 1, Verses 9–11 of the Bible that "what has been will be again, what has been done will be done again; there is nothing new under the sun." There is nothing that one can point to and say that it is new. Who knows, it was in existence but not remembered by past and future generations. Whenever we face plagues, they are unprecedented disasters for us. However, when we emerge from a plague, we forget the price and lessons we had paid over time.

With the rapid development of modern technology, many fields had entered new unmanned grounds from a historical perspective – environmental destruction, life science exploration, biomedicine research and development, and the development of intelligent technology. Human beings will always be unconsciously hypnotized by the unprecedented technological prosperity, indulge in prosperity brought by science and technology, and gradually lose their awe towards nature. While the development of science and technology benefited mankind, it led to us losing self-discipline and trying to continuously go beyond the borders of nature. And COVID-19 is a reflection of these acts, which saw some viruses in nature breaking through the immune barriers in humans and entering our society in the same manner.

Originally, we only required self-discipline to greatly minimize the damage caused by the virus. It can be said that in modern national governance, we are fully capable of strengthening our self-discipline. We only need to learn from China, Singapore, Japan, and other models to manage the citizens with appropriate self-discipline management methods. Similarly, we only need to wear a mask and exercise basic self-discipline such as maintaining social distancing, which can directly and effectively reduce the spread, infection, and fatality rate of the virus. However, we witnessed many people in many countries doing things their way despite the virus' severe threats. Many continued placing their hopes on technology and vaccines. This is not wrong. Every major plague in human history ultimately depended on the power of science and technology to be effectively controlled and resolved.

But technology is not omnipotent. We are both powerful and vulnerable in the face of nature. We seemed to be constantly exploring nature and using technology to broaden the boundaries of our existence while compressing the boundaries of nature. We forgot one of the simplest truths, which is that our knowledge of nature is still very limited. When parts of the ecological environment were destroyed, changed, and compressed by us, various unknown pathogens in nature would inevitably break through the boundaries between man and nature through various self-mutation methods, via the human immune barriers, and be used as the new host for their survival.

A large number of lives were lost in this pandemic due to excessive freedom and democracy in the West today, the United States being a classic example. The United States, a country established based on the Puritans, is a powerful country built on extreme self-discipline and a strong adventurous spirit. However, due to various types of arrogance, it gave up self-discipline in the name of democracy and freedom and abandoned the most basic way of protecting oneself – wearing a mask.

This major plague will inevitably end soon. However, if we cannot coexist in harmony with nature in a self-disciplined way, perhaps the next plague of a bigger scale is already waiting for us. In my view, the next pandemic would originate from the ocean. The following are the two main factors:

On the one hand, thousands of unknown viruses and bacteria live in the ocean itself. However, it has not been given due attention due to pollution, which already led to increasingly serious damage to the marine ecological environment. When the ocean bacteria's living space and environment were invaded and destroyed, they would mutate someday unknown to us, break through our barriers and use us as their new host.

On the other hand, the melting of glaciers caused by global warming triggered the return of an ancient unknown virus to the world. In the second half of 2020, a paper was published in the journal *Contemporary Biology*, where scientists discovered a 57,000-year-old prehistoric corpse in the permafrost of the Arctic. This is the best-preserved "mummy" of a wolf pup, to date. It can be said that the exceptional preservation of its body when it was discovered, alongside the ability to analyze the contents in its stomach, was due to the protection by the frozen soil layer. So, how was it discovered? It was the melting of the permafrost that led to the discovery of its body. Under the thick ice layers of the Antarctic Circle and the Arctic Circle, there was frozen soil that existed for a long time. This is highly dangerous because there may be many ancient biological bodies or germs buried within. These ancient unknown viruses were sealed under the thick glaciers with the help of God, but unfortunately, we promoted and accelerated global warming of the earth by leveraging various science and technology methods, resulting in the glaciers melting at an alarming rate and the possibility of these ancient, sealed germs being active amongst us in the near future.

The only way for us to effectively avoid the great plague is to respect, protect and maintain a harmonious relationship with nature. If we do not take a serious view on environmental protection, I can be sure that the price we pay for the next crisis will inevitably be greater than this plague. Self-discipline and nature are responsibilities that we must shoulder. We need the self-discipline to restrict our actions and co-exist harmoniously with nature. We should not think that technology can help us to break through natural barriers in the flourishing age of technology. For us, caring for nature is the best protection for our survival.

References

1. L. Bos. *Beijerinck's work on tobacco mosaic virus: historical context and legacy.* 1999, 354 (1383): 675–685.
2. Lily E. Kay. *W. M. Stanley's crystallization of the tobacco mosaic virus, 1930–1940.* 1986, 77 (3): 450–472.
3. Homer. *The Illad.* [M] Volume 1.
4. Zhao Tianen. *An overview on the history of leprosy in ancient China [J]. China Journal of Leprosy and Skin Diseases,* 2011, 27 (01): 73–74.
5. Michael Kulikowski. *Justinian's Flea: plague, empire, and the birth of Europe. 2007,* 35 (4): 148–148.
6. Gu Yingtai. *History of the Ming Major Events.* [M] Volume 78.
7. Boylston A W. *The origins of vaccination: myths and reality [J].* Journal of the Royal Society of Medicine, 2013, 106 (9): 351–354.
8. Katherine E Arden, Ian M Mackay. *Human rhinoviruses: coming in from the cold.* 2009, 1 (4): 913–920.
9. Briese Thomas, Renwick Neil, Venter Marietjie, et al. *Global distribution of novel rhinovirus genotype.* 2008, 14 (6): 944–7.
10. Andrew Moravcsik. *Pandemic 1918: Eyewitness accounts from the greatest medical holocaust in modern history.* 2019, 98 (2): 179–180.
11. Li YH, Chen SP. *Evolutionary history of Ebola virus [J].* Epide-miol Infect, 2014, 142 (6): 1138–1145.DOI: 10.1017/S0950268813002215.

12. Jiang Chenyang, Lian Xiaodong, Gao Ce, et al. *Distinct viral reservoirs in individuals with spontaneous control of HIV-1.* 2020, 585 (7824).
13. Gong Yue, Shi Zhixiang, Chen Jing, et al. *Current Status of Research on Coronavirus [J].* China Bitoechnology, 2020, 40 (Z1): 1–20.
14. Special Expert Group for Control of the Epidemic of Novel Coronavirus Pneumonia of the Chinese Preventive Medicine. *An update on the epidemiological characteristics of novel coronavirus pneumonia (COVID-19) [J].* Chinese Journal of Viral Diseases, 2020, 10 (02): 86–92.
15. Cleri DJ, Ricketi AJ, Vemaleo JR. *Severe Acute Respiratory Syndrome (SARS).* Infectious Disease Clinics of North America. 2010,24 (1): 175.
16. Zumla A, Hu DS, Perdman S. *Middle East respiratory syndrome.* Lancet, 2015, 386 (9997): 9954007.
17. Aiu I, Kojima K, Nakane M. *Transmission of severe acute respiratory syndrome. Emerging Infectious Diseases,* 2003, 9 (9): 11834184. [59]
18. Goh DLM, Chia KS, et al. *Secondary household transmission of SARS, Singapore.* Emerging Infectious Diseases, 2004, 10 (2): 232–234.
19. Gerald W. Volcheck. *How the immune system works.* 2001, 86 (3): 350–350.
20. Effros RB. *Roy Walford and the immunologic theory of aging.* Immune Ageing, 2005, 2 (1): 7.
21. Walford RL: *The Immunologic theory of aging [M].* Copenhagen: Munksgaard Press, 1969.
22. Myroslava Protsiv, Catherine Ley, Joanna Lankester, et al. *Decreasing human body temperature in the United States since the industrial revolution.* 2020, 9.
23. Lungato L, Gazarini ML, Paredes-Gamero EJ, et al. *Paradoxical sleep deprivation impairs mouse survival after infection with malaria parasites.* Malaria Journal, 2015, 14: 183.
24. Almanan Maha, Raynor Jana, Ogunsulire Ireti, et al. *IL-10-producing Tfh cells accumulate with age and link inflammation with age-related immune suppression.* 2020, 6 (31): eabb0806-eabb0806.
25. Franceschi C, Capri M, Monti D, et al. *Inflammaging and anti-inflammaging: a systemic perspective on aging and longevity emerged from studies in humans [J].* Mech Aging Dev, 2007, 128: 92–105.
26. Montecino-Rodriguez E,Berent-Maoz B, Dorshkind K. *Causes, consequences, and reversal of immune system aging [J].* J Clin Invest, 2013, 123: 958–965.
27. Nishimura H, Nose M, Hiai H, et al. *Development of lupus-like autoimmune diseases by disruption of the PD-1 gene encoding an ITIM motif-carrying immunoreceptor [J].* Immunity, 1999, 11 (2): 141–151.
28. Dong H, Zhu G, Tamada K, et al. *B7-H1, a third member of the B7family, co-stimulates T-cell proliferation and interleukin-10 secretion [J].* Nature Medicine, 1999, 5 (12): 1365–1369.
29. Nagavendra Kommineni, Palpandi Pandi, et al. *Antibody drug conjugates: development, characterization, and regulatory considerations.* Polym Adv Technol, 2019; 1–17, DOI: 10.1002/pat.4789.
30. Zhang Baihong, Yue Hongyun. *Efficacy predictors of tumor immunotherapy [J].* Cancer Progress, 2019, 17 (01): 7–10, 61.
31. Sang Woo Park, Daniel M. Cornforth, Jonathan Dushoff, et al. *The time scale of asymptomatic transmission affects estimates of epidemic potential in the COVID-19 outbreak.* 2020, 31.

32. Edward M. Behrens Gary A. Koretzky. *Cytokine Storm Syndrome: Looking Toward the Precision Medicine Era.* https://onlinelibrary.wiley.com/doi/full/10.1002/art.40071
33. Jennifer R. Tisoncik, Marcus J. Korth, et al. *Into the Eye of the Cytokine Storm.* A Microbiol Mol Biol Rev. 2012 Mar; 76 (1): 16–32. https://mmbr.asm.org/content/76/1/16
34. Tung Thanh Le, Zacharias Andreadakis, Arun Kumar, et al. *The COVID-19 vaccine development landscape.* 2020, 19 (5): 305–306.
35. Lee Wen Shi, Wheatley Adam K, Kent Stephen J, et al. *Antibody-dependent enhancement and SARS-CoV-2 vaccines and therapies.* 2020.
36. Janssen Vaccines & Prevention B.V. *A Randomized, Double-blind, Placebo-controlled Phase 3 Study to Assess the Efficacy and Safety of Ad26.COV2.S for the Prevention of SARS-CoV-2-mediated COVID-19 in Adults Aged 18 Years and Older [R].* 2020.
37. Xu, J., et al., *Orchitis: A Complication of Severe Acute Respiratory Syndrome (SARS) 1.* Biology of Reproduction, 2006. 74 (2): p. 410–416.
38. Wadman, M., et al. *How does coronavirus kill? Clinicians tracea ferocious rampage through the body, from brain to toes.* Science 2020.
39. Wells A U. *The revised ATS/ERS/JRS/ALAT diag-nostic criteria for idiopathic pulmonary fibrosis (IPF)-practical implications [J].* Respir Res,2013,14 (Suppll): S2-S6.
40. Yong-Zhen Zhang, Edward C. Holmes. *A Genomic Perspective on the Origin and Emergence of SARS-CoV-2. 2020,* 181 (2): 223–227.
41. Xiaolu Tang, Changcheng Wu, Xiang Li, Yuhe Song, Xinmin Yao, Xinkai Wu, Yuange Duan, Hong Zhang, Yirong Wang, Zhaohui Qian, Jie Cui, Jian Lu. *On the origin and continuing evolution of SARS-CoV-2 [J].* National Science Review, 2020, 7 (06): 1012–1023.
42. Xiao Kangpeng, Zhai Junqiong, Feng Yaoyu, et al. *Isolation of SARS-CoV-2-related coronavirus from Malayan pangolins.* 2020, 583 (7815): 286–289.
43. Song Xinming. *Development of the Epidemiological Transition Theory – Formation and Development of Epidemiological Theory in Population Change [J].* Institute of Population Research, 2003 (06): 52–58.
44. Fan, C., et al. *ACE2 Expression in kidney and testis may cause kidney and testis damage after 2019-ncov infection.* medRxiv, 2020: p.2020.02.12.20022418.
45. Marinho Paula M, Marcos Allexya A A, Romano André C, et al. *Retinal findings in patients with COVID-19.* 2020.
46. Jonathan P Rogers, Edward Chesney, Dominic Oliver, et al. *Psychiatric and neuropsychiatric presentations associated with severe coronavirus infections: a systematic review and meta-analysis with comparison to the COVID-19 pandemic.* 2020, 7 (7): 611–627.
47. Li Xueying. *Cognitive theory and cognitive behavioral therapy of PTSD.* Chinese Journal of Clinical Psychology, 1999, 7 (2): 125–128.
48. Chen Gen. *Digital Twins [M].* 2019.
49. MIT Technology Review. *Israeli scientists invented reusable masks, and USB connected mobile phone chargers can eliminate viruses.* [EB/OL]. http://mapp.mittrchina.com/mittrchina_h5/app/shareArticleInfo/shareArticleInfo.html?uid=26103&ccid=2587

50. Harriet Constable. *Pneumonia epidemic: What would a city that can resist infectious diseases look like.* [EB/OL]. https://www.bbc.com/zhongwen/simp/amp/world-52529304?__twitter_impression=true
51. Qiu Jian, Li Jing, Mao Suling, Li Yi. *Research framework of urban vulnerability and planning response under major epidemics [J/OL].* Urban Planning: 1–9 [2020-10-20]. http://kns.cnki.net/kcms/detail/11.2378.TU.20200909.1447.008.html
52. Ifeng Finance. Long Yongtu. *Someone started discussing about "De-sinicization", and we must remain highly vigilant.* [EB/OL].
https://finance.ifeng.com/c/7wKHwBqo2uO
53. Fudan-Pingan Research Institute for Macroeconomy. *Small and medium-sized enterprises under the epidemic.* [R]. 2020-5-8.
54. Zhu Chenbo. *Discussion on the development of China's aviation industry under the impact of the epidemic [J].* Civil Aviation Management, 2020 (06): 22–25.
55. Li Ning. *Evolution of the Hollywood film industry model (2009–2019) [J].* Contemporary Film, 2020 (04): 112–120.
56. Future Think Tank. *Digital Foresight Research: Digital New Infrastructure, Digital Ecology, Digital Economy [R].* 2020-6-7.
57. Zou Songlin. *From a bad example to a role model: What does South Korea rely on in fighting the epidemic without a lockdown? [J].* China Economic Weekly, 2020 (06): 99–102.
58. Deep Knowledge Group. *COVID-19 Regional Safety Assessment (200 Regions) [EB/OL].* https://www.dkv.global/covid-safety-assesment-200-regions
59. Zhang Zedong, Liu Hong. *Gradual decision-making and governance capabilities: Taking Singapore's fight against COVID-19 as an example [J].* Hubei Social Sciences, 2020 (08): 42–51.
60. Gu Linsheng, Ren Jiehao. *Japan quickly responded and implemented measures In the face of the epidemic [J].* China Emergency Management, 2020 (02): 52–54.
61. The Editors. *Dying in a Leadership Vacuum.* 2020, 383 (15): 1479–1480.
62. Ran Ran. *Anti-intellect, populism: US epidemic governance under Trump's leadership [J].* Journal of the Central Institute of Socialism, 2020 (03): 53–60.
63. (U.S.) *Adler. Six Concepts [M].* Beijing: Sanlian Bookstore, 1989.16.
64. *Ancient Greek and Roman Philosophy [M].* Beijing: Sanlian Bookstore,1957.49.
65. Meng Peiyuan. *The subjective thinking of Chinese philosophy [M].* Beijing: People's Press. 1993.121.
66. Larry M. Bartels. *Unequal Democracy: The Political Economy of the New Gilded Age [M].* Princeton: Princeton University Press, 2008: 112.
67. James D. Hunter. *Culture Wars: The Struggle to Define America [M].* New York: Basic Books, 1991.
68. Morris P.Fiorina et al. *Culture War: The Myth of a Polarized America [M].* New York: Pearson Longman, 2005.
69. Liu Mingming, Liu Yi. *The institutional roots of the US's epidemic situation [J].* Frontline, 2020 (08): 20–23.
70. Liu Baihui, *Taking reference from U.S. Health Power Division and Transfer Payment System,* Sub

National Fiscal Research, No.8, 2016.

71. 2 K. J. Arrow. *Uncertainty and the Welfare Economics of Medical Care,* The American Economic Review, Vol. 53 (5), 1963.
72. Lai Shengqiang, Tang Xuemei. *Research on the Influence of Information Emotion on the Spreading of Internet Rumors [J].* Journal of Information 2016, 35 (01): 116–121.
73. Li Yingtao. *Analysis of the "vulnerabilities" and "vulnerable groups" in the global COVID-19 pandemic [J].* Journal of International Political Studies, 2020, 41 (03): 208–229, 260.
74. Coordination of Humanitarian Affairs, (2021, May 1). *COVID-19 Global Humanitarian Response Plan: United Nations Coordination Appeal (April-December 2020).* https://www.unocha.org/sites/unocha/files/Global-Humanitariar-Response-Plan-COVID-19.pdf
75. *Protection of human rights protection in special times – The fight against the epidemic is inseparable from the consciousness of a community of shared destiny,* People's Daily, 3rd Edition, April 16, 2020.
76. Li Yinghong. *On the Return of Mutual Subjectivity between Human and Nature [J].* Journal of China University of Mining & Technology (Social Sciences), 2020, 22 (04): 90–100.

Index

ABOUT THE AUTHOR

Mr. Kevin Chen, or Chen Gen, is a renowned science and technology writer and scholar, visiting scholar at Columbia University, postdoctoral scholar at University of Cambridge, and an invited course professor at Peking University. He has served as special commentator and columnist for *The People's Daily*, CCTV, China Business Network, SINA, NetEase, and many other media outlets. He has published several monographs involving numerous do-mains, including finance, science and technology, real estate, medical treatment, and industrial design. He has currently taken up residence in Hong Kong.